Whether recounting nights spent searching for moths amid the heather or relating an autumn dedicated to the perfect blue of Clifden Nonpareil, this boy can write!

If moths mean nothing to you, openin
street into an unexpected party. Here
jolly, generous, kind-hearted host, Jam
moth intoxication!

Patrick Barkham, author of *The Butterfly Isles*

Gloriously uplifting, hilariously eccentric; a big warm hug of a book written straight from the heart. Moths at their most inspiring, nature writing at its finest.

Helen Pilcher, author of *Life Changing*

James Lowen confesses his love affair with some of Britain's most overlooked creatures – and, in doing so, reveals the wonder of moths. A delightful book, packed with passion and fascinating detail.

Stephen Moss, naturalist and author

A charming book.

John Ingham, *Daily Express*

A great read.

Nigel Marven, wildlife presenter and naturalist

This is a book full of enthusiasm and erudition.

Adrian Spalding, *Atropos*

Lowen brings a charm and wit to these close encounters [with moths], making them personal and intimate, and a delight to read.

Richard Jones, Royal Entomological Society

Written with craft and class [...] The ride is as mad as a moth's meanders.

Dominic Couzens, author and journalist

Written by someone who so ably conveys his passion, *Much Ado About Mothing* is an enthralling 20-chapter celebration of these winged insects. Accompanied by his abiding enthusiasm and wonder, Lowen's writing is entertaining, packed with descriptive prose and fascinating facts about his quarry.

Josh Jones, *Birdwatch*

Charming and awe-inspiring. Whether you love or loathe moths, this book is for you.

Kate Bradbury, author of *The Bumblebee Flies Anyway*

With prose as rich and velvety as a Black Rustic's wings, in Much Ado About Mothing James Lowen shines a welcome light into the hidden world of Britain's moths [...] their stories are remarkable and, in this delicious book, Lowen serves them with the relish they deserve.

Jon Dunn, author of *Orchid Summer*

James has a great way with words and brings moths to life, from the tiniest micro to the largest macro.

Dave Grundy, *Comma*

This journey introduces the reader to many astonishing species [...] a thoroughly enjoyable read.

Ashleigh Whiffin, *BBC Wildlife*

A very good read. Highly enjoyable.

Mark Avery, conservationist and co-founder of Wild Justice

A Note on the Author

James Lowen is an award-winning author specialising in travel and natural history, two of his 13 books receiving the accolade of Travel Guidebook of the Year. After living in South America and working variously as an Antarctic tour guide and environmental policymaker, James now combines writing articles for UK newspapers and magazines with raising his daughter.

In his forties, having long disdained moths, the scales fell from James's eyes, inspiring him to write both *Much Ado About Mothing*, which was longlisted for the Wainwright Prize for nature writing and chosen as *Country Life* magazine's 'Book of the Week', and *British Moths: A Gateway Guide* (Bloomsbury Wildlife, 2021).

@JLowenWildlife

www.jameslowen.com

MUCH ADO ABOUT MOTHING

A year intoxicated by Britain's rare and remarkable moths

James Lowen

BLOOMSBURY WILDLIFE
LONDON • OXFORD • NEW YORK • NEW DELHI • SYDNEY

BLOOMSBURY WILDLIFE
Bloomsbury Publishing Plc
50 Bedford Square, London, WC1B 3DP, UK
29 Earlsfort Terrace, Dublin 2, Ireland

First published in the United Kingdom 2021. Paperback edition 2023.

A catalogue record for this book is available from the British Library

ISBN: PB: 978-1-4729-6698-8; HB: 978-1-4729-6697-1; eBook: 978-1-4729-6699-5

2 4 6 8 10 9 7 5 3 1

Typeset in Bembo Std by Deanta Global Publishing Services, Chennai, India
Printed and bound in Great Britain by CPI Group (UK) Ltd, Croydon, CR0 4YY

To Maya, Sharon and Will
In memoriam Douglas Boyes (1996–2021)

Contents

Note on names

For larger moths, I use the vernacular names published in the *Field Guide to the Moths of Great Britain and Ireland* (written by Paul Waring and Martin Townsend; Bloomsbury, 2017). For smaller moths, I follow the *Field Guide to the Micromoths of Great Britain and Ireland* (written by Phil Sterling and Mark Parsons; Bloomsbury, 2018). Micromoths are universally known by their scientific names, so I follow suit. However, where Sterling and Parsons include common names, I use those preferentially. On the odd occasion, I also take advantage of common names for micromoths that are in wide usage but do not feature in Sterling and Parsons.

Prologue

I remember the date precisely: 7 July 2012. It was the day my life changed for ever.

It wasn't the day I got married. Nor was it when our daughter was born. It *did* involve a female, but it wasn't a girl or woman. It brought a *coup de foudre*, a wholly unexpected buckling of the knees, an unanticipated thumping of the heart. For she was arrestingly beautiful. And she was a moth.

To be precise, she was a Poplar Hawk-moth. She was sweetly furry, verging on velvety. This winged wonder was gawkily angular – Darth Vader's TIE Fighter in animal form. She was wrapped in silver, banded with bronze and grizzled with iron filings. She was enduringly placid, only once flashing me a warning with fiery spots on her underwings. And she was huge and glorious and utterly wrong.

Because moths – I knew this for a fact – were small, brown and dull. At best, they were uninteresting and usually irritating. These eerie creatures of the night remained invisible yet destroyed my suits. Moths were malign phantoms from another realm. My attitude to moths was not, however, constant. Some days I hated them. At other times they merely stultified me. Occasionally I was even indifferent.

How I disappointed my nature-loving buddies! Most, like me, had long expanded their wildlife horizons from birds and mammals to butterflies and dragonflies. Many had jumped the ditch dividing things that move from things that grow, becoming increasingly enamoured by the unabashed ostentation of orchids. Again I was among their number. But not when it came to moths. 'When,' I once asked sulkily, 'will I be so bored of other wildlife that I have to resort to getting kicks from moths?'

For these poor relations of butterflies were (still) small, brown and dull; (still) uninteresting and often irritating. Several friends begged to disagree; they had already seen the light. Andy sang hymns of praise to the hundreds of types of moth that routinely visited his garden. Martin had already inked onto his Christmas list a copy of a new field guide to Britain's smaller (for which, read often seriously tiny) moths. Mark sought to effect my Damascene conversion to his invertebrate religion with tales of moths that mimicked wasps to evade predation and others – veritable vagabonds – that migrated here from Africa. Evolutionary marvels and the allure of the rare were astute buttons for Mark to press. But still he failed. I remained resolutely anti-moth.

Until my friend James Hunter cracked me on that fateful July day. Then a trainer for a biotech company who was invariably clad in a baseball cap and hoodie, James caught the train from Dartford to Blackheath and wandered uphill to our 1960s terrace. He was joining me on an excursion to a shady Beech hangar in the Chilterns to admire a graceful and near-extinct orchid, the Red Helleborine.

I knew that James was a moth-lover or 'moth-er' – naturalist parlance that avoids confusion with the female parent at the expense of pronunciation and ugly punctuation. James spoke readily of tigers and footmen, of waves and wainscots. He had branched out from seeking to attract moths in his garden with the wan, blueish strip light of an actinic bulb. He now thought nothing of running a generator overnight to power super-bright mercury-vapour lights that illuminated Kent's downland darkness. This exploratory zeal fired my spirit of adventure, which was desperate to be rekindled following several years living in South America. But still moths did not captivate me.

Until, that is, James opened a scruffy rucksack and extracted a Tupperware pot the size of a doorstep sandwich. Something large, grey and winged lurked inside.

'Mate,' he said, 'take a look at this. It's a Poplar Hawk-moth.'

Grudgingly, I looked. And was immediately, shockingly and overwhelmingly smitten. My life has never been the same since.

Admittedly, it took eighteen months for me to really get going with 'mothing'. Make no bones about it, moths can seem a daunting group of animals to wrap one's head around. Many people, myself included, find it hard enough to learn Britain's butterflies, dragonflies and orchids – groups that comprise sixty or fewer different species. Moths are a different ballpark. There are roughly forty times more types of moth in Britain than there are butterflies – some 2,500 species. At the outset, the variety is simply bewildering. Granted, many are striking, colourful and pleasingly easy to identify, but many others look very similar to one another. Picking up an identification guide to moths for the first time risks making your eyes bleed: on some pages, all the moths look the same. And if that wasn't challenging enough, perhaps two-thirds of Britain's species are 'micromoths', tiny wisps of life whose appreciation demands a hand lens.

Then there's the kit involved. Unlike butterflies, the vast majority of moths do not flaunt themselves in the sun or pose on flowers. Although 250 or so moths readily choose to fly by day, that still leaves 90 per cent of Britain's species to emerge only under the cover of darkness. To see them, you need to play canny by attracting them to light – for which moths seemingly have the most bizarre fascination. At its simplest, this can mean leaving a bathroom light on and the window open. A less hit-and-miss approach is to invest in a powerful light source and connect this to an open-mouthed 'trap' stacked with egg-boxes on which moths rest, unharmed, after being attracted to or bewildered by the unexpected ground-level 'moon'. After identifying and photographing the moths, the moth-er returns them to the wild, hiding them deep in vegetation away from the beady eyes of hungry birds. The financial outlay for a moth trap is typically a couple of hundred quid, but it is worth every penny.

Learning to put the right name to moths is like starting to ride a bike: initially impossible but, with practice, pleasingly straightforward. The joy that catching and identifying moths can bring proves unbridled, instructive and revelatory. I would never have guessed that three hundred types of moths – almost double the collective total of species of butterflies, dragonflies and orchids in the whole of Britain – would visit my modest London garden, otherwise unwitnessed. For home is where every moth-er's experience begins.

Moths make for gloriously lazy wildlife watching. No need to travel, no need to rise early, no need to seek out. Let moths come to you. Pootle out to your trap after breakfast and see what the night has divulged. My daughter and I were routinely astonished and swiftly became addicted. It transpired that moths were not always small and brown. Quite the opposite. Many rendered some butterflies tawdry, others dullards. It became apparent that moths could be brightly coloured, classily patterned and surprisingly large. Some species were all three: Maya fell head over heels for the sizeable, boldly striped and candy-pink Elephant Hawk-moth.

This was particularly gratifying, for we caught scores of them. Here is the other thing about trapping moths: you catch *so many*. Wander around your garden on a sunny summer day, and you might – if there are suitable nectar sources – see a handful of butterflies. Open your moth trap after a summer night, and you can expect two orders of magnitude more than that. On one remarkable morning during my first proper summer of mothing, there were 2,800 of the winged wonders sitting peaceably inside. *Two thousand eight hundred*. Roughly 2,500 of them were a single tiny species – Horse-chestnut Leaf-miner (*Cameraria ohridella*) – but that still left three hundred other types. The sheer volume of moths that are out there, unseen and unappreciated, beggars belief.

That Horse-chestnut Leaf-miner – a beautiful tangerine-toned species, banded with silver – is an interesting one. It was unknown in Britain before 2002, but has spread far and wide across England and Wales, becoming common wherever Horse-chestnut trees grow. This is not, however, a zero-to-hero story but one of zero-to-alleged-villain. Forestry Research, Britain's public body responsible for tree-related research, classifies Horse-chestnut Leaf-miner as a pest. The moth's caterpillars munch away the tree's leaves, causing them to discolour before prematurely falling to the ground. In truth, this does not appear to impoverish the tree's health. But that, for public body and general public alike, is beside the point. This moth is a pest, and pests must be persecuted.

By lazy association, all moths are vexatious. This one chomps leaves, but others devour our clothes and carpets. And we *really* don't like that. Ergo all moths are evil. Worse, they do their ghoulish business unseen, so we have little opportunity to make their acquaintance, to demystify them, to befriend them. Or they fly erratically at us – seemingly in attack. Do they bite, we fret? Then throw in the caterpillars of Brown-tail Moths, which – should you believe the more scaremongering of UK tabloids – can blind you if their hair (somehow) touches your eye, and moths become dangerous as well as destructive.

Pilloried, slighted and vilified, moths are Mother Nature's bad boys. Butterflies, those poster children of the insect world, have it easy. But here's a thing: moths and butterflies are one and the same. All are members of the order Lepidoptera (from the Greek: *lepis* for scale and *pteron* for wing). In evolutionary terms, butterflies actually nestle as a grouping *within* moths: the 'family tree' sees moths on each side of butterflies. The latter just happen to fly by day and to have club-shaped antennae rather than feathery, ribbed or saw-edged feelers. Other than that, there is little consistent difference – certainly nothing justifying such vastly different public perception. Everyone loves butterflies; everyone seemingly hates moths. By any rational measure, the discrepancy in opinion is untenable.

Fortunately, not quite everyone hates moths. Take my daughter, for instance. Turning four during our inaugural summer of mothing, she could barely read, let alone be indoctrinated by tabloid sensationalism. Maya's eyes were open to new experiences, unfettered by prejudice. The wild child loved what she saw and, as kids are supremely tactile, adored what she could hold. In two decades of watching butterflies, I can count on one hand the times I have enjoyed direct, physical contact with one: a Green-veined White wiggling onto my outstretched palm; a Green Hairstreak treating my finger as a branch; a Purple Emperor lapping up my sweat through its whirligig proboscis. In contrast, most moths are unfazed extroverts that sit placidly while being examined, and some seemingly even hanker after the human touch. As a device for inspiring kids about nature, there is nothing better or more readily accessible.

Towards the end of that inaugural summer of mothing, we moved house – to a new county, a new life. From a claustrophobically narrow, three-storey house in The Smoke, we relocated to an airy, sprawling bungalow on the outskirts of Norwich, our garden quadrupling in size. Maya could not wrap her head around the move. She had somehow gleaned that we would abandon not just the building but also all its contents – including her toys and our moth trap. It took aeons to reassure her that this was not the case. Yet, evidently enamoured by her time with the winged insects, she was distraught that 'Daddy's moths' would not come with us. Don't worry so, I soothed, there will be even more moths where we are going. Maya's face brightened. A smile supplanted tears. Life was going to be OK.

Our Norfolk garden proved pretty good for moths. Perhaps not as startling, at first blush, as London but still pretty good. We took our time making their acquaintance. We discovered how the types of moths ebbed and flowed with the passing of the seasons. We interrogated our visitors' life stories – marvelling at moths with a prowess for camouflage that would make a chameleon proud (the Buff-tip is a dead ringer for a snapped twig) and at those with antifreeze in their blood (enabling the well-named December Moth to survive early winter chills). We deciphered the connection between caterpillar foodplants in our garden and the adult moths we caught. We gawped at insects whose migratory feats made birds look like wimps, such as the titchy Diamond-back Moths (*Plutella xylostella*) that arrived from Russia by the thousand.

These last two features – foodplants and migration – illustrate the compelling dichotomy of moths. For foodplant, read *place*. For migration, read *displacement*. Place and displacement, location and dislocation. Arguably, no British animals are more entwined with place than are moths. One species, Fisher's Estuarine Moth, was long thought not to stray beyond ten metres from the plant where it hatched. Another, New Forest Burnet, now resides only on a single, remote sea cliff in west Scotland, exhibiting blatant disregard for its name.

With their finicky ecological needs, both species have a fragile grasp on survival; extinction could be just one washed-out summer away. As such, they and many, many other moths tell vibrant stories about the wonderful, varied landscapes they inhabit and how we have transformed them.

In 1976 a single Devon trap operated by the Rothamsted Insect Survey – a nationwide network of sites that provides decades of standardised data about invertebrate populations – caught 4,681 individual moths in one night. I can recall sitting in my parents' car later that decade as we weaved along the same county's narrow, hedge-shrouded lanes. Returning to our holiday cottage from an evening meal at the local pub, the balmy summer air would thicken with moths and other winged insects before splattering an untimely end on the windscreen. This phenomenon, enshrined in the title of Michael McCarthy's environmental treatise *The Moth Snowstorm*, is no more. Within a generation, the moths have vanished. Driving back from a nearby village inn, with Maya dozing in the back, I spot just three. Official statistics, compiled by the wildlife charity Butterfly Conservation, back up this subjective impression. The populations of many British species have bellyflopped by as much as 80 per cent over thirty years. Their mass disappearance is trying to tell us something about how we are treating our countryside.

But enough already about place. What about displacement? Here too, moths are winged messengers for the state of our planet. As we have seen, the Horse-chestnut Leaf-miners that thronged my London trap were unknown in Britain twenty years ago. Last century, seeing this flaming-orange, burning-eyed insect would have been a moth-er's dream. Then they arrived, colonised, prospered, spread, prospered some more and spread some more. And they are far from alone. Whereas Britain's contingent of butterflies is pretty much fixed – with only sporadic attempts at establishment by continental residents such as Queen of Spain Fritillary and Long-tailed Blue – each year sees a fair group of moths recorded here for the first time. Some are one-off migrants. Others – Sombre Brocade, Tree-lichen Beauty, Gypsy Moth and their ilk – have swiftly become resident. Still others are tentatively following suit. All bear witness to a changing climate and to the interconnectivity of a globalised world.

After we fled London in favour of a quieter, simpler and smaller life in Norfolk, the other thing that struck me about moths was that I was no longer alone. When in London, the nearest friend who ran a moth trap lived ten miles distant, in Kent. I basically had nobody to share sightings with or to guide my learning. Around Norwich, however, there was a thriving community of moth-ers. Gradually, we compared notes on who had caught what, where and when. Tentatively, we answered one another's identification conundrums. And, when we trapped something exciting or exotic, we started to invite our brothers- and sisters-in-arms to come and see it.

Knowledge, companionship and horizons expanded. I realised there was a moth world beyond my garden. Initially it comprised fellow moth-ers' homes. Then I learned that nature reserves and Butterfly Conservation organised 'moth mornings'. Aiming to enthuse people about moths, they invited the public to attend the opening of traps sited in landscapes as diverse as reedbeds and rivers, heathland and fen. I had seen four hundred different types of moths so far, but here were ways to potentially double that. Entire pages of the field guide, where the paintings were as much myth as moth, suddenly became available for first-hand scrutiny. Blinkers had fallen from my eyes.

This got me thinking: there comes a point in any new undertaking where you yearn to take a big step forward; it is human nature to push boundaries, to explore further, to go beyond. I found myself craving more moths, different moths, moths elsewhere, moths *from* elsewhere. I had heard of birdwatchers doing 'big years', striving to see three hundred-plus types of bird within twelve months. I had read books by nature writers who had taken things a step further, aiming to cram all of Britain's dragonflies, butterflies, orchids, rare animals or rare plants into a single year or so. My heart was telling me with passion that it was time to devote a similar chunk of my life to moths.

But what should The Quest comprise? There are relatively few butterflies, dragonflies and orchids in Britain, so seeing them all in a year is a realistic ambition, but only a fool would attempt a clean sweep of Britain's two thousand-plus moths in a single year. There are more than enough species for an entire lifetime – and with newbies added every year, the full house is a moving target. I am a novice moth-er: seeing my remaining 1,800 new moths in a year is unimaginable. Yet targeting just, say, fifty moths would be a betrayal of moths' stellar diversity, their evolutionary prowess. Instead, I decided to see an ambitious and varied suite of scarce and special creatures, each with a tale worth telling. I elected to pursue quality over quantity – prioritising Britain's rarest and most remarkable moths. I would target masters of camouflage and deception, conservation success stories and failures, new arrivals and elusive species, the protected and the pestilent. I would revel in the boldly colourful, but also find beauty in the tiny. Above all, I would deepen my understanding of why moths are brilliant – and then shout the answers from the rooftops.

Seeing my desired 120-ish non-garden-dwelling species would involve travelling to many places, landscapes and habitats that would be new to me, but also enable me to view familiar locations through a fresh filter. My quest would take me from England's southwestern tip to northern Scotland. I would camp out in ancient woodlands and idle across sunny heathlands. I would slosh through bogs and yomp up mountains. Along the way, I intended to retrace paths followed by Victorian collectors who provided the baseline for our understanding of British moths while making a decent living out of trading prized specimens. I would even chance my arm at rediscovering species feared extinct.

The counterweight to such travels would be our suburban garden. Here, at a relaxed pace and with contrasting ease, my daughter and I would track the seasons through moths, taking joy in the uncommon beauty of common creatures. For perhaps the greatest virtue of moths is their accessibility. Moths are everywhere, but above all they are *here*.

I was under no illusions that such weighty travels would be easy. Quite the opposite – even discounting the impact on family and finances. Unlike orchids and other plants, moths move, so grid references rarely help. Unlike butterflies, most moths hide by day so need to be sought in the pitch-black. To avoid thousand-mile failures, I would need up-to-date information, assessments of the impact of unseasonable weather, specialist equipment, patience, sharp eyes and luck – plus a bottomless pit of help and goodwill on the part of numerous experts, whether amateur moth-ers or professional conservationists.

This in itself was propitious, for my curiosity was increasingly piqued not solely by moths but by the people who choose to watch, trap, count, study, photograph and protect them. I wanted to understand why they do so – particularly when society's prevailing attitude is one of disgust for the object of their love. I wanted to learn how and why moth-ers get into moths. I wanted to understand how moths impact their lives and how moths make them *feel*. Initial sleuthing suggested there might be moth-ers every bit as obsessive as the bird-chasing twitchers who sometimes make the news. Through the filter of moths, what might I learn about the concept of obsession, both within other people and myself as the year unfolded? That initiatory Poplar Hawk-moth – huge, glorious and utterly wrong – would, I suspected, have a lot to answer for.

1

The winter garden – and beyond

Norfolk
January–March

It's an unexpected, blessed cure for a New Year's Day hangover: the rejuvenation of Alka-Seltzer, but in the form of a moth.

Struggling from bed, I grump about before slouching off in search of restorative eggs and juice. But before even passing through the supermarket's sliding doors, I am revived. On one of its towering plate-glass windows, gripping Spiderman-like to the pane's smoothness and transfixed by the light from within, sits my inaugural moth of the year. In normal circumstances, I might overlook it as a dirty thumbprint. But this year's circumstances are anything but normal. Throughout my quest, moths will be my new brown.

Unperturbed, the creature allows me to approach its rounded, dust-grey equilateral triangle. I get close enough to examine the gentle undulations of light and dark playing across its inch-wide wings. It is a male Winter Moth. Although not a rare species – far from it – it is remarkable, joyous and headache-alleviating.

The duality of moths underpins their attraction. In essence, they are out-of-scope creatures, inhabiting a world other to our own – that of darkness. Shine a light, however, and they materialise. The emptiness is revealed to be replete, and your garden confesses its inner nature reserve.

There is no universally accepted explanation for moths' draw to flame. One theory holds that lamps are mistaken for a celestial cue used in orientation – a low-hanging moon, in other words. Another suggests that night-flying insects are dazzled; another that they are overstimulated. It doesn't help our understanding that light affects different moths in different ways: some spiral around the source; others crash into it. Some settle at a distance; others ignore it. Nor does it help that some moths seem indifferent to particular wavelengths of ultraviolet light, yet are attracted to others. Mercury-vapour (MV) bulbs attract five times as many moths as high-pressure sodium streetlamps. Actinic bulbs seduce winter-flying moths more than summer-active types. And the jury's out on what effect new-fangled LED lights have. But put some kind of light in your garden and moths will come.

So I do. On 151 dusks across the year, I fire up my MV or actinic traps, curious to learn what visits the family hearth while we snore. The dazzling MV demands a berth in the secluded front garden, whereas the less obtrusive actinic will not disturb neighbours out back. Neither garden is large: the front a badminton court, at best; the back a quarter of that. Our Norwich bungalow is distinctly suburban – on an anonymous 1920s housing estate, its previously husbanded vegetation has been left to rewild in the absence of chemicals. Although beloved, it is hardly special.

But moths get everywhere. Even an inner-city location such as London's Natural History Museum garden has racked up six hundred species – including one new for the UK. You don't even need a garden. In the absence of outside space, two flat-dwelling friends independently installed a moth trap in their respective bedrooms, opening windows to allow moths in. Each caught hundreds of species, their totals limited only by extinguishing the light come bedtime. With moths, anything is possible anywhere.

It is this realisation that hooks hundreds, perhaps thousands, of people each year – all newcomers to the hobby. One recent convert is Pauline Hogg, a friend from Nottinghamshire. 'It's a whole secret world waiting to be discovered,' she enthuses. After attending some moth events, Pauline 'began wondering what was around my own garden' and so dangled an MV bulb over a tablecloth heaped with egg-boxes. She quickly became smitten. 'This hobby will absorb me for a very long time,' she concludes.

Norwich teacher Annie Farthing agrees. She treasures garden mothing as a way to connect her family with nature. 'On a weekend morning, rummaging through the night's catch is a lovely accompaniment to coffee and toast,' she smiles. Her husband Justin considers it 'a great pastime for wage slaves,' adding proudly that their six-year-old daughter, Ruby, 'would like to sleep by day and stay up all night with the moth trap.' Moths' surprising equanimity optimises interest for hands-on kids; most moths sit still, even allowing themselves to be held. Everyone loves the ceremony of mothing too: turning off the electricity; the nervous steps to the trap, with pots and notebook in hand; the careful removal of the bulb and housing; the unveiling of the Perspex lid or glass vanes that prevent the moths escaping; then, heart pounding, the reverential examination of both sides of every egg-tray. 'Every time is like Christmas morning,' Annie says.

That said, winter mothing is challenging. For a start, I feel like hibernating rather than looking for insects. More importantly, most moths are ectotherms, relying on external sources of heat to warm their bodies into activity. Heat and winter make rare bedfellows. Accordingly, seeing garden moths in the year's opening months is hard work. I catch only single figures – if any at all – until an unseasonably balmy spell at February's close.

But at least there *are* moths in winter. The odd lethargic Peacock or sun-awoken Red Admiral aside, no self-respecting butterfly would be seen dead during the parky months. Although lovely, butterflies are lightweights. Moths are where it is at. Take, for instance, that inoffensive window-shopping Winter Moth: unremarkable on the outside, but quite the opposite inside. Like some other winter-flying geometers – moths whose shape most closely resembles butterflies – this creature produces a kind of antifreeze, enabling activity at low temperatures.

A larger relative, Pale Brindled Beauty, conjures a similar trick. Maya spots the year's first: 'Dada, what's this big moth on the kitchen window?' It is straighter-winged than the Winter Moth and half as large again. Its bulbous head swells into a furry thorax-cum-winter-muff, while generously feathered antennae splay perpendicular. In his book *Enjoying Moths*, Scottish lepidopterist Roy Leverton recalls discovering a Pale Brindled Beauty encased in thick ice after landing in a pool that subsequently froze. Leverton chipped out the moth then thawed it. At dusk, the moth flew away, apparently unharmed. The antifreeze done good.

Winter-flying moths are assuredly tough creatures. We catch several Dotted Border, a softly triangular geometer that ripples with the buffs and browns of a desiccated leaf. The disguise becomes apparent when a friend shares a photo of one resting on fallen oak leaves. It takes me sixty seconds to discern the moth, before marvelling at its dusting of black spots that match the leaves' mould spores. Dotted Border rises further in my estimation when I encounter one transfixed in the same position eight days apart. Temporary dormancy saves energy – another survival technique. It becomes a firm favourite when I spot seven plastered to ice-licked walls at Glenshee Ski Centre in the snow-quilted Scottish Highlands. I am shivering below my thermals; this is one hardy insect.

The three moths mentioned – along with other geometers that grace our garden, such as Early Moth, Mottled Umber and Spring Usher – share another extreme adaptation to winter cold. To save energy, the species' females have forsaken the ability to fly. They crawl everywhere, relying on the fully-winged males to travel to their service. A female Early Moth in a friend's garden appears more woodlouse than moth, all stumpy body and abbreviations for wings.

As its name suggests, the Oak Beauty we catch assumes the mantle of winter's best-looking moth – the catwalk creature in a season of relative nondescripts. The species emerges early this year: a friend even catches one in January. This large, hirsute moth boasts boldly textured wings emblazoned with shards of chocolate and froths of cappuccino. Its nearest relative, Peppered Moth, is famed for having evolved a dark-coloured morph that optimised camouflage from predators during the soot-tinged, air-polluted Industrial Revolution. In the Netherlands, Oak Beauty exhibits an all-black form for similar reasons, yet oddly the same genetic quirk has not wheedled into British populations.

Emerging in late autumn, Mottled Umber just about persists into the New Year. More individuals than usual do so this January, meaning we encounter several under streetlights plus more in our garden trap. In its most colourful form, with its flair for chestnut and creamy diagonals, Mottled Umber aspires to be an Oak Beauty – albeit a slighter, more fragile variety. Even so, I appreciate why Justin Farthing is taken aback when reaching into the fridge at his workplace to discover a Mottled Umber huddling inside. 'It must have been brought in with the milk,' he surmises, wonder inscribing his face.

As I compile the year's first garden-moth records, ready to send to the county moth recorder where they can be analysed, a double-edged thought punctuates the spreadsheet drudgery. The first realisation is the opportunity furnished by moths to reckon the year's passage by its natural calendar, rather than feeling constrained by Gregorian dates. In an age characterised by nature-deficit disorder, I seek liberation through sensitising myself to moths' seasonality.

The rub comes in the emerging disconnect between the garden-dwellers' names and established concepts of season. Even without leaving home, Maya and I can discern apparent impacts of climate change. Because winter flight is unusual in moths, several cold-habituated species are named after months or seasons. But the warming climate is disrupting natural timing, a concept known as phenological change. The emergence dates of most British butterflies have advanced significantly over the past thirty years. The same is true of moths. Across four hundred species analysed by Butterfly Conservation, mean flight dates now rank nearly five days earlier than during the 1970s.

The Spring Usher has become as much a January moth as one of February's tail. Far fewer Early Moths are on the wing in March nowadays compared to four decades previously; perhaps it should be renamed Ever-earlier Moth. March Moth still broadly adheres to its name, but April sees many fewer airborne than last century. How long will it be before Winter Moth bids for a new moniker?

2

Cats, tracks and caves

Norfolk, Derbyshire and Dorset
January

What came first, the moth or the egg? Or – given the complex life cycle that characterises Lepidoptera – the moth, egg, caterpillar or pupa? The now-familiar but ever-remarkable metamorphosis of moths and butterflies from crawling caterpillar to winged adult was not grasped until the seventeenth century. Caterpillars, then usually called 'worms', were thought to emerge spontaneously from dewy leaves and to be an entity separate from moths. Once understood, the insects' metamorphosis became deployed as a metaphor for human spiritual growth or even resurrection – the soul being liberated from the pseudo-sarcophagus of the chrysalis.

Winter is a quiet period for moth-ers, perfect for assimilating how their quarry *becomes*. This comprises searching for what are collectively called the 'early stages': the egg (also known as ovum), caterpillar (larva) or pupa (chrysalis). In Britain, relatively few adult moths (imago) are active in winter. Adults' comparative paucity reflects the season's low temperatures (meaning many foodplants are unavailable), wind and rain. Survival is the priority. Each species has its preferred strategy. More than half of British species spend winter as caterpillars; nearly a third remain in the snugness of a chrysalis, often buried underground. The remaining sixth splits evenly between eggs and adults.

I commence my personal quest for these early stages at what an individual moth would consider its own beginning: the egg. One January night, I join Ben Lewis, cheery warden at the RSPB's Strumpshaw Fen reserve, and talented bug-hunter Phil Saunders to search Alder-rich woodland in Norfolk's Yare Valley. After half an hour, Ben beckons us towards the lichen-green gnarl of a hefty trunk. Sprawling outwards from a bark crevasse are scores of tiny, tawny ring-doughnuts packed in a tight rectangle. They are eggs, each less than a millimetre across and collectively hugged by a grey, silky web. The threads are remnants of the cocoon from which a female Vapourer herself first crawled. Being flightless – as are the females of thirteen other British macromoths – laying eggs nearby makes sense, and the web provides modest protection from parasitoid wasps. Across just a few square centimetres I count upwards of 260 ova, which she will have laid during late summer or autumn. She will most probably have completed her life's work with brutal efficiency, enticing a male to mate within three hours of becoming an adult herself. By now, she is long dead. Moths do not do childcare.

Each species trades off the number of eggs against their size. Large ova harbour abundant food stores, giving larvae the best start in life, but there's insufficient space inside a female for hundreds. Little eggs produce small, vulnerable caterpillars but can be carried in proliferation, producing more offspring. Most species seek a compromise between quality and quantity.

Timing is also nip and tuck. If you're an egg during winter, you're best off staying that way until your foodplant is available. Hatching too early – before leaf burst – means insufficient food. But hatch too late, and the sustenance may have gone. Timing is key because larvae are essentially eating machines. Among the species whose adults lack mouthparts and thus cannot feed, larvae are *essential* eating machines, responsible for securing all the energy and nutrients the moth will need until the end of its adult existence. These are Very Hungry Caterpillars.

Moths that emerge as adults in summer typically accelerate through their life stages, overwintering as caterpillars. I could probably find several juvenile life forms by rummaging in my garden. But for something more worthwhile, I help Butterfly Conservation's Sharon Hearle survey Lunar Yellow Underwing caterpillars in Norfolk's Breckland. Covering nearly four hundred square miles, the Brecks' sandy heathland just about convinces as Britain's answer to the Mongolian steppe. The area harbours a rare and precious assemblage of moths that will coax me here time and time again in summer. This includes Britain's biggest population of tonight's target species. During the first sixteen years of this century, agricultural intensification deprived the moth of a quarter of its national range. It is now confined to the Brecks, coastal Suffolk and the chalk downland marrying Wiltshire and Hampshire.

As I depart Norwich, the sky is leaden, laden even. But as I reach Cranwich Camp – a former labour camp and home to a Royal House Artillery regiment – the cuttlefish-ink clouds clear and the rain threat abates. The cooling air is speared by a Woodcock, all fat-belly urgency and take-no-prisoners bill. Since Cranwich's demolition, nature has reasserted itself across a square of sandy grassland. Somehow thronging with rare life forms, it has become a Site of Special Scientific Interest. Not that people uniformly respect that status. At the site entrance, three mattresses have been fly-tipped on a remnant concrete foundation around which one of Britain's rarest plants, Proliferous Pink, will flower come July. Cranwich is a strange place but an ecologically valuable one.

Which justifies the strange activity underway in a rectangle of predominantly bare soil. Bums facing the stars, four people are crawling on hands and knees, peering just above ground level. Alongside Sharon and me are experienced local moth-er Graham Geen and Abbie Rix, a Master's student. Keen lepidopterists love a larval hunt – indeed, the technique was fundamental before moth traps became widely used – so it surprises me that I am not the only Lunar Yellow Underwing caterpillar virgin. Graham is too.

Sharon tells us that excavators created the bare ground seven years previously to help Spanish Catchfly, another rare plant, and scarce insects.

'We're trying to replicate survey methods from previous decades, to see how things have changed,' she explains. 'As there's four of us, we'll do a timed count for fifteen minutes. The caterpillars hatched in September and don't pupate until April. They are small at the moment – perhaps two centimetres long. They hide by day then, at night, crawl up grass tufts to dry out. To see them, lie on your tummy, beaming your torch ahead.'

As instructed, we slither and shine. My sciatica yowls in protest. Sharon soon spots the first Lunar Yellow

Underwing caterpillar. She is right: it is teeny, but also rather tubby with its dull olive tones fettered by creamy stitches. We pinpoint just two more throughout the remaining twelve minutes.

Bemusement seasons Sharon's disappointment. 'We've been finding loads elsewhere – up to eighty-nine. But here the vegetation is really short. I wonder if they've been munched by grazing animals.' Grazing, if carefully managed, is a useful conservation tool; Sharon's scenario would be an unfortunate unintended consequence. Sadly, it's time for me to leave, but visiting Cranwich has whetted my appetite for Breckland moths and I will be back. The other three depart to survey another site: more tummying along the ground, more bums in the air, and hopefully many more teeny cats.

'You've got to be a little bit nuts to do moths,' Sharon concludes by way of farewell. I sense that our paths will cross again.

Two weeks later, as I find myself sliding and stumbling down a scree slope that drapes above a sheer sea cliff, Sharon's words assume greater resonance. I am following Mark Parsons, who is celebrating twenty years as Butterfly Conservation's head honcho for moths, seemingly towards certain injury. Sure-footed, rugged-looking and insatiably inquisitive, Mark has enticed me to a cliff of shattered limestone on the sheltered, east-facing coast of Dorset's Isle of Portland. We are searching for one of the world's rarest moths.

'The entire global range of *Eudarcia richardsoni* – common name Richardson's Case-bearer – was long thought to comprise just two sites in Dorset, including Portland,' says Mark. 'But it was found in the Swiss Alps eight years ago, and possibly Croatia since.' It probably occurs elsewhere too, Mark thinks, but nobody has found it yet. Understandably, as it transpires. The adult moth is four millimetres long, barely the size of a fruit fly. It's a smidgeon of life, moth dust. But we're not seeking the adult. We're not quite looking for the egg, caterpillar or pupa either. We're rooting around for this micromoth's larval case.

'It rather resembles a mouldy grain of rice,' Mark advises. For someone whose gateway moth was a glam Poplar Hawk-moth, this furnishes an unsettling search image. Mark explains that the caterpillar dwells inside a portable housing – a live-in sleeping bag – tethered to rock. From within this protective layer, formed from tiny rock fragments and lichen particles, the larva nibbles algae and possibly insect detritus trapped in spider-webs. The youngster stays like this for two years until pupating into adulthood.

Mark indicates an accumulation of scree comprising chunks about the dimensions of my heavily gloved hand. I should turn over each rock fragment and scrutinise the underside for a camouflaged larval case. After twenty minutes, a couple of layers down, I spot two pale, fusty objects shaped like mouse poo. Tentatively, I call Mark over, expecting to make a fool of myself.

'Well done!' he exclaims. 'That's *richardsoni*.' In this guise, the moth is no looker: 'mouldy grain of rice' seems a generous description. But it's the most excitingly anomalous moth that I have yet seen. It's also thrillingly rare.

The adult moth was first found by C.R. Digby near Swanage in 1882 but misidentified. Larval cases were then discovered by Nelson Richardson here at Portland in 1894. Richardson reared the invertebrate peculiarity through to adulthood, identifying it as a sister species of *Eudarcia*. It took until 1900 for the Rt Hon. Lord Walsingham to realise the moth was new to science and formally describe it. Politician and landowner, this typically upper-class Victorian entomologist boasted one of the world's most bountiful collections of micromoths, some quarter of a million specimens. Walsingham was a giant of his entomological era, authoritative on the tiniest of creatures.

Although he modestly objects to being described as an expert, Mark is indubitably among those who have assumed Walsingham's mantle. Together with Butterfly Conservation colleague Phil Sterling – a man whose enthusiasm catapults you into excited optimism – Mark has done much to make micromoths accessible to laypeople. Alongside genius artist Richard Lewington, in 2012 the pair wrote the first modern field guide to these tiny creatures. The book has rapidly fuelled interest in their existence, distribution and conservation.

Mark and Phil have also done more than anyone to understand and help Richardson's Case-bearer. Ninety-eight years after Digby's encounter, they rediscovered the moth at Swanage. Since 2013 they have been monitoring it on Portland as part of Dorset Wildlife Trust's attempts to stop invasive cotoneaster plants swathing the Isle's limestone. Like Sharon Hearle in the Brecks, Mark standardises surveys through timed counts.

This year's instalment, conducted shortly after my visit, revealed only ten larval cases across eight sampling sites, the lowest-ever count. Because the habitat doesn't seem to have changed, Mark is not yet panicking about this remarkable, über-rare moth. *Yet.*

Before we leave, Mark makes time to introduce me to other moths' larval life. He splits apart the desiccated seed head of a Carline Thistle. Inside is a tiny white grub, the caterpillar of *Metzneria aestivella*, which feeds solely on this spiky plant. Mark then gestures towards a dense white web tenting a bramble bush. I recognise this one: the caterpillar commune of Brown-tail Moths. First causing alarm in 1780s London, this species has become a scourge of local authorities because of the itchy rash the larvae's tiny hairs can provoke. It prompts us to discuss the widely held perception that some moths are 'pests', a word to which Mark stridently objects.

He picks a green bramble leaf. It is decorated with a long, wiggly line whose creaminess, when held against the light, proves translucent. This, Mark explains, is a leaf mine, the intricate gallery left when the caterpillar of a micromoth has nibbled between a leaf's protective cuticles. Feeding between leaf surfaces is an evolutionary tactic that avoids the plant's chemical defences but radically constrains the size of the caterpillar and resultant moth. Adult leaf-miner moths are tiny, difficult to detect and harder to identify, so a dedicated clade of moth-ers instead focuses attention on the distinctively wandering tracks of their caterpillars – a niche practice to detect niche moths. From the tapeworm form of its riverine meander and the location of its 'frass' (a posh word for moth poo), Mark identifies this one as *Stigmella aurella*.

It's all way above my pay grade, but Mark is undeterred. Under a shady overhang, he plucks a leaf from Pellitory-of-the-wall, a herb clambering between cliff fissures. It displays an irregular whitish blotch – the mine of a *Cosmopterix pulchrimella* caterpillar. This moth was not known in Britain before 2001 when Mark discovered it in Dorset, confirming his identification with the help of Phil. Definitely not experts, these two.

Flippancy aside, the study of leaf mines has greatly deepened understanding of the diversity and distribution of Britain's moths. It provides an efficient means of keeping tabs on our rapidly changing environment. The first record for Norfolk of one leaf-miner, *Ectoedemia heringella*, comprised several million mines on Holm Oak trees at Thetford – a sudden, dramatic arrival.

It doesn't have to be leaf mines either; some caterpillars leave even more curious forensic evidence. One dark evening a month later, Norfolk moth-er Keith Kerr furtively passes me some contraband from his car boot. Squinting at the apparently innocuous pine cone, I see a two-millimetre-wide hole in the tip of one of its seeds, as if pierced by a compass. To Keith's delight, this proves the egress – and thus local existence – of *Cydia conicolana*, a nationally scarce micromoth that otherwise lives undetected high in pine trees. Keith reckons he had checked a thousand pine cones without success, then his dad found a hole in the very first cone he plucked off the ground.

'I was half-sweary, half-happy,' Keith admits. 'The species was new for the ten-kilometre square – our own "dot on the map". It doesn't matter that we didn't see the moth itself. The point is that it *had* been there and that *we* had found it.'

'Ah, bagworms. Yes, that way lies madness.'

Phil Sterling should know. He has devoted a considerable proportion of recent decades to studying what are arguably Britain's oddest moths, the twenty species brigaded in the family Psychidae. Bagworms see the larval-case element of Richardson's Case-bearer and raise it. Bagworm caterpillars build cases ('bags') from fragments of plant material, attach the structure to a tree or foliage, and live out their larval existence within. So far, so very case-bearer, you might think. Yet from here on, bagwormery gets wacky.

I see two species of bagworm during my interrogation of moths' early stages. Initially, I find *Psyche casta* attached to a toilet-block wall, then spot them affixed to other substrates including, bizarrely, my hall window. The species' larval case (*casta*) appears to be six wooden splinters extracted from a child's thumb and stuck together in a whelk-like whirl. It's hard to credit the structure as containing animal as well as plant. But the one on my window moves six inches one night, and – triffids aside – plants don't do that.

Madder still is *Luffia lapidella*, whose larva lives within what resembles a tiny rabbit-food pellet with a tapered tip. My inaugural encounter with this space oddity involves a congregation of two hundred shambling slowly across the lichen-smeared boundary wall of a Norfolk housing estate. The next, unfathomably, clings to the metal postbox that stands guard outside our front gate. What she expects to eat there is beyond me. I say 'she' with confidence as, in almost all of Britain, only female *Luffia* occur. They reproduce parthenogenetically – laying eggs that have no need to meet sperm in order to create embryos and, eventually, new moths. Almost uniquely among British moths, this is virgin-birth territory. Perhaps even weirder, the grub-like female's entire life cycle takes place inside her larval case: pupation, 'emergence' as a wingless adult and egg-laying. *Luffia* challenges the very concept of moth.

Sadly, I fail to come across Britain's oddest bagworms, *Pachythelia villosella* and *Acanthopsyche atra*. The latter has intoxicated Phil, who spent some time on the Dorset heaths trying to find it, then penned a scientific paper summarising what he had fathomed about its life history. Like the two bagworms I did manage to see, *Acanthopsyche* and *Pachythelia* females are wingless and grub-like. Unlike that pair, however, *Acanthopsyche* and *Pachythelia* put to remarkable use their uncanny resemblance to a pale, scale-free maggot. Swathes of moths have evolved characteristics designed to evade predators. In stark contrast, upon exiting their larval case, females of these two species cravenly wiggle their nakedness, apparently striving to be gobbled up by bird or reptile. Experiments have shown that larvae can hatch from eggs ejected in the predators' droppings, suggesting this to be an intentional strategy to disperse progeny. Getting eaten enables the female's offspring to conquer new terrain. A mother's love knows no bounds.

All this exposure to moths' early stages has me pining for adults – and musing about the ways *they* spend winter. I've already seen a few species in my garden, of course. But what about those that live through winter without actively experiencing it? What about the hibernators?

Britain's most famous hibernating moth is the Herald, so named for the adult's broad, tabard-shaped wings emblazoned with fiery livery. In autumn, when the moth emerges, it could pass for a decaying leaf in its favoured broad-leaved woodland. But that serves little purpose in winter, when leaves and tree are decoupled. As they will not breed until spring, Heralds spend winter in hiding, finding shelter in dark, cool places – from caves to cellars, culverts to outbuildings – then flipping the off-switch until spring. Suitable locations are comparatively scarce so dormant moths often share a dormitory, whiling away the colder months in a huddled congregation – a triumph of Heralds.

I receive invitations to see Heralds hibernating in a cave, a Second World War bunker, a bat hibernaculum, an eighteenth-century ice house and a tramway tunnel. Will Soar guides me around the tunnel, which lies near the Derbyshire village where he grew up. In his early thirties, Will's auburn-tinged beard scrubs over a youthful face. Now Norwich-based, he earns his keep alerting birdwatchers to the presence and absence of rare birds. In his teens, a birdwatching friend showed him the mothing light.

The tunnel lies in the former estate village of Ticknall, whose heritage stretches back to the Domesday Book, which inscribes it as Tichenhalle. Ticknall's heyday spanned the turn of the nineteenth century, when the brick-making and tile industries levied an inexhaustible demand for local lime, thereby supporting a population thrice that of today. An industrial relict of that era – a tunnel that ushered a tramway from lime yards to the Ashby canal – provides our destination on a blustery, saturated winter afternoon.

As we squelch into the 120-metre-long tunnel, the air stills and chills. Although not quite going underground, our search for buried treasure unfurls amid darkness. Wetness oozes through the limestone roof, hinting at mini-stalactites and crawling with spiders. There is no massed gathering of Heralds, but singletons and braces dot the two-metre-high ceiling. Our buried treasure totals a meaning-of-life forty-two moths plus half a dozen Peacock butterflies.

Our visit is just a one-off. East Lothian-based Katty Baird, however, passes much of her life in the dark depths where Heralds sleep. Together with Mark Cubitt, Katty runs the Hibernating Heralds project, a citizen-science initiative that engages people about Scotland's moths while expanding understanding of the whereabouts and lifestyle of Heralds and Tissues, a fellow moth-troglodyte. The project has encountered amazing success. From December 2016 to spring 2019, Katty and Mark received 1,290 records of 8,444 Heralds from 362 locations – ten times more hibernating sites than were previously known across the whole of Scotland. The pair have discovered that Heralds spend the majority of their life in the dark. Most arrive in September or October, not usually moving an inch until departing in April. Colour-marking individual moths unveiled one particularly slothful soul that arrived on 28 August and stayed immobile for 264 days.

The project appeals to the inquisitive scientist in Katty. 'I like to ask and answer questions,' she admits. 'Surprisingly little is known about the Herald's ecology or behaviour – that honour tends to be preserved for pests or rarities. So it's satisfying to continually add nuggets of information to the bigger picture.' The buzz Katty and Mark derive from the quest stretches well beyond the animals themselves. Reaching sites often involves negotiating 'tightly packed contours, water, gorse or all three,' she explains. Her thirst for adventure is clear, as is the means that moths provide to that end: 'Mark and I have explored some fascinating places that I never would have known existed had it not been for Heralds.'

3

The spring garden – leaves, twigs and bird craps

Norfolk
March–May

Of all seasons, spring is the most coquettish towards a moth-er. Daytime warmth and leaf burst induce anticipation that can be scotched upon nightfall, when the residual heat has an unfettered escape route through clear skies and moths decline the invitation to fly. Traps stay empty, spirits sag, moth-ers mope.

But moths may also give when you least expect it. On 22 March, our garden trap breaks through the hundred-moth mark on its earliest-ever calendar date. Even so, two-thirds are relative nondescripts called Quakers – cannon fodder for those who consider moths small, brown and boring. Catches remain unseasonably strong until April Fool's Day, which laughs cruelly as it ushers in lepidopteran despondency. The month's first five nights do not warrant the electricity invested, averaging just one moth apiece. Despite trapping more frequently than in March, we catch barely one-third the total number of moths.

Just when you assume things cannot get worse, May frosts up. Norwich friends record blank nights redolent of January's doldrums. 'Enthusiasm tempered,' one grumbles. 'Dismantling the trap until it thaws,' moans another. Only during the last third of the month do counts pick up. A tenderly sunny morning with 132 moths of forty-two species gladdens the soul, even if our time is spent rehousing scores of chitin-coated cockchafers that sprawl comatose around the trap like morning-after partygoers.

This burgeoning variety also furnishes a rich vein of life for Maya and me to appreciate and ponder. Each morning before breakfast, bubble-wrapped with bird noise, we enthusiastically riffle through the egg-trays before secreting their temporary denizens into the undergrowth where they can resume life after the brief trap-bound hiatus. The species we discover provide stories – of ecology and evolution, camouflage and conservation – that ferry us through many a school run.

We admire the fluffily feline Pale Tussock as it luxuriates, forelegs outstretched, then marvel at its ability to produce in-flight sound through organs called tymbals, thereby advertising its distastefulness to bats. The shag-pile furriness of a male Muslin Moth pleases us greatly; it has grey, mad-professor hair and an orange buttercup-chin. I explain that the male flies by night yet the female by day – and wonder how they ever coincide to procreate. We career around the garden to enact the charge of the Light Brocade, an attractive stained-glass window of a moth. Iron Prominent, meanwhile, is a nightclub bouncer: stocky and swarthy, rusty patches creeping across gunmetal wings. An Orange Footman is shaped rather like a pumpkin seed. During my lifetime it has become an astonishing hundred-fold commoner, a beneficiary of air-quality improvements that have enabled its lichen foodstuff to flourish. We gawp at Spectacle, an unexpectedly natty goth-moth arrayed in grey and black that has an upstanding gingery Mohican and, when regarded front on, wears John Lennon specs on what passes for its forehead. 'That moth can't be real, Dada,' Maya exclaims. 'Those eyes are in the wrong place.'

Another personal favourite, Brindled Beauty, proves worryingly scarce this year. The grouchy appearance of this large, grizzled moth belies a peaceable, friendly demeanour that enchants Maya. Its woollen greyness recalls the fusty grandpa overcoats I wore in my teenage Dr Marten years. Nationwide data collected since 1970 – among 25 million records analysed by Butterfly Conservation – reveal both good news and bad. Brindled Beauty is becoming more widespread in Britain, its distribution extending by nearly a quarter in forty-six years. Simultaneously, however, this same species is getting rarer, its long-term abundance slumping by more than three-quarters since 1970. Moth-ers are catching Brindled Beauty over a wider area, but we're getting many fewer of them. The reasons are unclear but, for moths more widely, habitat degradation and fragmentation plus agricultural intensification top the list, with climate change following close behind – pushing species north, adversely affecting species with a single generation early in the year and spinning ecological relationships out of sync. Were moths not drawn to the flame of ultraviolet light, I tell Maya, we would understand disturbingly little of their world and how we are impacting it.

One morning, the season's first Peppered Moth provides chat for the entire trip to school. This world-famous, large, long-winged insect exists in two main colour forms: one white with black speckles, and the other wholly coal-black. In Britain, the former predominated until the Industrial Revolution, at which point the melanistic morph came to the fore, ostensibly because it was less visible to predators on soot-layered trees. Dark phases actually accreted in a hundred species of moth, but Peppered stole the headlines. Eminent Victorian entomologist James Tutt posited that darkness thrived through natural selection. Since 1970, our less dirty air has eroded the advantages of sooty colouration; with the gene producing freckles now offering better camouflage on cleaner, lichen-caked trunks, pale morphs have resurged.

Inspired by my eight-year-old's grasp of Darwinism before breakfast, I chance my arm with the less well-known, scientifically controversial story. In the 1950s, under the tutelage of celebrated ecological geneticist and lepidopteran giant E.B. (Henry) Ford, Bernard Kettlewell conducted experiments that proved the natural-selection theory, canonising the moth in scientific textbooks. Critics – notably Judith Hooper, in her 2002 book *Of Moths and Men* – picked apart Kettlewell's work, accusing him of fabricating evidence. It took seven years of study by a third entomologist, Michael Majerus, to vindicate Kettlewell's research and Tutt's original postulation about camouflage, bird predation and the rise and fall of melanistic moths.

'Dada, I'm hungry.'

My digression has fallen on deaf ears. Darwinism before school is overkill.

Ill-judged parental enthusiasm notwithstanding, our discussion piques Maya's interest in camouflaged insects. Through our garden, we investigate creatures that go to remarkable lengths to elude detection by moth-munchers.

Three species provide individualistic takes on the notion of a withered brown leaf. Scalloped Hook-tip holds its brown-lined beige wings in an elevated tent shape; their trailing edge undulating like a crumpled leaf. The leading edge of a Lilac Beauty's richly toned forewings pleats back on itself, resembling a desiccated bract. Maya's eye lingers on the moth's pink shimmer; a gentle smile creases her face. My wife Sharon joins us to admire the season's inaugural Angle Shades, which is flying earlier in the year than in the 1970s – a sign of our warming times. This moth is sumptuous: streamlined wings, each folded in on itself and crinkled at the rear; a symmetrical interplay of army-surplus olive-brown plus rose, violet and brass. Try spotting that in a tangle-wood.

From leaves, we move up-tree to twigs. Several chunky moths compete for the accolade of 'best stick'. The woodchip-like Pale Pinion narrowly loses a podium spot to Chamomile Shark, a localised moth of chalky grasslands. As its name suggests, this creature's hunched back is topped with a Jaws-like fin, while its eyes, remarkably, are veiled with lashes. Taking silver is a fellow 'shark', Mullein. Its snapped-twig impression surpasses that of Chamomile Shark by dint of parallel go-faster stripes including a blackish ridgeback. Mullein also wins on sneakiness. Long confined to its namesake plants, a transient growth on disturbed ground, the moth's caterpillars have developed more easily sated taste buds. Mullein now munches buddleia, that rampant butterfly-bush of brownfield and suburban gardens. We can expect more in the coming years.

Neat though they are, this trio are variations on a theme and so they are unambiguously trumped by the mother of all garden moths, Buff-tip. Whichever way it is aligned along a slender branch, it loses itself seamlessly by dint of being buff at both tips. A flattened, clotted-cream head is ringed with umber, like bark around a cleanly broken branch. At the end of silver-grey wings lies a broad slash of beige, as if a branch has been snapped off at an angle. Viewed from either end, the disguise is unsurpassable – and when looked at top-down, delicate black marks appear to pick out the face of a sleeping person. Sharon rightly swoons each time we catch one: 'That moth is the coolest thing in the world!' Then I ill-advisedly venture that it also resembles a fag butt, and she biffs me in outrage.

Twigs are not the only tree parts that we espy in our moth trap. Waved Umber is a piece of flaking bark, Early Grey and May Highflyer both winged patches of lichen. Among the most luscious of spring moths, with its fusion of russet, gold and silver, Pine Beauty merges nicely into a pine trunk but disappears even more effectively into a budding pine cone. Most giggle-worthy of all, however – particularly if you are a child (or her dad) – are the bird-poo moths. Splashes of black and white excrement, these moths hide in plain sight. After all, no self-respecting Blue Tit would ever eat a Great Tit's dropping, would it? And thereby Chinese Character, Lime-speck Pug and various piebald tortrix micromoths live another night.

Buff-tip was a prelude to moths that elicit a difference of opinion. Maya is adamant that the Brimstone – a lemon-yellow wonder, lying with its wings open – is a butterfly rather than a moth. Sharon queries my assertion that an Early Thorn – sitting with its wings snapped shut, just like a butterfly – is a moth. A Red-green Carpet poses likewise, then lies its wings flat to reveal a sensual mosaic of scarlet flirting with pea-green.

This triumvirate prompt discussion about what differentiates a moth from a butterfly. The rightful answer – and one that will distress moth-haters everywhere – is: 'not much'. Taxonomically, all are scale-winged creatures sitting within the order Lepidoptera, the six butterfly families actually nestling within the 120-ish families of moths. Their life cycle – egg, caterpillar, pupa, adult – is identical. For sure, butterflies have clubbed antennae – but so do burnet moths. Butterflies tend to be brightly coloured – but so are the three moths under dispute, for a start. Moths tend to have plumper bodies than butterflies, granted, but the divide is blurred. Nor do butterflies have exclusive solar rights; hundreds of British moths fly by day while Red Admiral, a butterfly, routinely migrates by night. Linguistically, English is an outlier in differentiating between moths and butterflies. In French, there are butterflies (*papillons*) and night-butterflies (*papillons de nuit*). German (*Tagfalter* and *Nachtfalter*) and Dutch (*vlinder* and *nachtvlinder*) adopt similar approaches. Moths are simply butterflies with bad PR.

With that in mind, one balmy late-March afternoon, Maya and I decide to try and entice to the garden a day-flying moth that – glamorous of appearance and imperious of demeanour – effortlessly puts to shame every single British butterfly. The conditions are perfect for worshipping Emperor Moths: warm and sunny, with a soft breeze. We need the air to be mobile so that it wafts sex scent towards libidinal males. In a peculiar practice common among moth-ers, we have hung a brick-coloured rubber bung from our washing line, letting it exude the synthetically produced pheromones of a female. Although undetectable to us, molecules from this lure will disperse far and wide. Once a male catches the scent on his fulsome, finely plumose antennae – from as far as two miles away, some reckon – he will surmise that an Empress is 'calling' her willingness to mate. By zigzagging across the plume, he will pinpoint the source then arrow in towards it. Or, in this case, towards us.

Sure enough, just as we are digesting our Saturday fish and chips, an orange fury blazes over the rooftop into the airspace above our patio. Frenetically the Emperor circles the lure, then alights on a colour photocopy of an Empress that I have pinned to a branch out of curiosity. To spare him the frustration of a dry hump, I net him. Once Britain's sole silk moth has stopped smoking, we can fully appreciate his splendour. The size of a Painted Lady butterfly, a leonine mane divides wings worthy of adulation. Atop a grey background suffused with scarlet, and between two sets of bold lines – one curving, the other undulating – sit two startling owl eyes, RAF roundels formed from wing scales. As I stroke the moth's thorax, he strains his upperwings apart, revealing a second set of enraged eyes upon his blazing hindwings. He then swings a round on the branch to face me. If anything, his underside is even more fearsome, four eyes fixing us here too, but the hindwings now washed in blood. The sultan of spring is in charge.

4

(What's the story) Kentish Glory?

Lancashire, Highland and Aberdeenshire
April

It's mid-April; time to travel. Well before silly o'clock on a Sunday, I fire up the Quattro for a 500-mile day trip. Distance matters not when you're chasing beauty. Or, in this instance, chasing Belted Beauty.

Beside me in the passenger seat that he will occupy for the next six months is Will Soar. Will makes a natural wingman for this year of craziness: he knows loads about moths, lives only a mile from me and can flex shifts to travel midweek. Keith Kerr fills the back seat. Having aspired to play rugby professionally until injury struck, Keith is massive. With his tattoos, buzz cut and biceps bigger than Will's thighs, he shatters any preconception that moth-ers are uniformly weedy nerds.

Keith ascribes his interest in wildlife – particularly minibeasts – to a childhood accompanying his dad on fishing trips. Like me, Keith recently upped sticks from London, urban stresses having deprived him of ready contact with nature. He couples a deep understanding of moths with an inquiring mind and a relentless devotion to pushing the boundaries of knowledge ('improbable does not mean impossible,' he says). I quickly admire this impressive, generous man.

Mid-morning, we reach the Lancashire seaside town of Heysham. It's an unprepossessing place, and we park at Potts' Corner beneath the rank ugliness of a nuclear power station. To our west a near-interminable expanse of sand sprawls to a slither of distant sea. To our south oozes a mile of saltmarsh, where short-turfed land is incised with the negative space of viscous channels. Above, the wind canters through a vast sky.

Wrapping up, we join a group of twenty-five volunteer surveyors being marshalled by local naturalist Steve Palmer, who welcomes us to 'the annual survey of England's only viable population of Belted Beauty'. A colony of this moth, Steve explains, was discovered here in 2002, following three isolated records across the previous twenty years. It was a major surprise. Belted Beauty had lost many sites since the 1920s, and just two other contemporary outposts were known across England and Wales. Moreover, like the species's stronghold on Hebridean machair, these remaining locations lay on sand dunes above the tideline. The Potts' Corner moths, in contrast, inhabit sandy saltmarsh that is inundated by tides higher than 9.5 metres.

'I've been interested in Belted Beauty ever since their discovery here,' Steve confesses. Nobody cares more deeply about this rare moth than him. Every week between early March and early June, he walks transects across the saltmarsh to count them: Belted Beauty is active by day as well as night. Monitoring ground temperatures through dataloggers buried at the depth where Belted Beauty pupates, Steve and former academic Alan Bedford are investigating potential correlations between ground temperature, tides and emergence periods. Intriguingly, spring tides might warm the earth, helping Belted Beauty to emerge. Steve also lobbies against harmful infrastructural development on this Site of Special Scientific Interest, persuading a wind-farm company to bury power cables beneath saltmarsh rather than gouging it with trenches.

Once a year, Steve arranges for citizen scientists to thoroughly survey the whole site. 'Volunteers help massively,' he says. Some are regulars. They include the Morris family: mum Nina plus teenage sons Jack and Josh. I've known Nina and her husband Pete for twenty years and have heard that the boys are incipient wildlife superstars.

This year, Steve first spotted a male Belted Beauty on 21 March. Twenty-five days later, I worry that we are surveying too late – and if we *are* too late, the year's first major excursion will fail, setting a forlorn tone for the coming months – but Steve knows better. The species has a protracted emergence period, so he is confident that we will encounter some. 'We've only once drawn a blank,' he says. 'But let's spread out to increase chances – and holler if you find one.'

Happily, it takes the razor eyes of youth barely five seconds to spot the first Belted Beauty as Jack calls 'Got one' just three metres ahead of me. From my height above ground, it resembles a pale splash of gull poo. I drop to the lawn of short fescue grass, the better to inspect. The moth is an unexpectedly diminutive male. His head and thorax ruff with a generous, diffusely striped mane. His wings swirl with cappuccino – frothy milk dusted with chocolate waves and veined with rich espresso.

That a moth possesses wings would not normally be notable. All butterflies fly; the clue is in the word. But moths differ. Belted Beauty is among the fourteen British moths where only the males can fly. We see this for ourselves five minutes later when Keith spots a she-moth laying tiny fluorescent-green eggs at well-spaced intervals along a desiccated tussock-grass cane. Unrecognisable as a moth, she is a visual oddity – a woodlouse wrapped in a frayed woollen scarf. The female's entire world is constrained by her capacity to crawl. Evolutionarily speaking, this seems bizarre. Wings facilitate finding a mate, conquering new land, escaping from predators. Whether reducing them to stumps (brachyptery), narrowing them (stenopterism) or forsaking them entirely (apterism), why abandon such useful structures? Yet in isolated landforms, such as oases, where branching out may lead to a hostile environment, natural selection favours staying put. If a female does not *need* to fly, she can redirect energy from strengthening thoracic muscles or powering flight into producing more or richer eggs.

Even accepting such logic, the UK distribution of Belted Beauty remains anomalous. How can a flightless moth live on islands? One theory is that it is an interglacial relic that trundled across western Scotland at times of lower sea levels. Another explanation surmises that females inadvertently hitch lifts on salt-washed rafts of debris, floating the seas until making landfall on the Hebrides. Both hypotheses feel improbable. 'But not impossible,' Keith grins.

Jack and Josh continue finding males and the odd female. Some cling to the leeward side of cluttered vegetation. Other males slob on open turf, unbothered by wind. One female has clambered up some shrivelled, tide-jettisoned seaweed. She has gained height, we reckon, to 'call' to males – announcing her wherewithal to mate. We wish her luck. She will need to breed before next weekend's particularly high tide, when her flightlessness risks rendering her a sitting but non-swimming duck.

After two hours, we regroup to play Belted Beauty bingo, shouting out our counts. Keeping a running total, Steve reaches fifty-six males and thirty females. 'That's middling, perhaps a bit low,' he observes. 'One year we had 1,500.' Steve's transects suggest a short-term decline from 2009 to 2016, although he cautions against reading too much into such a small dataset. That said, the moth hasn't been seen elsewhere in England since 2010, nor in Wales since 2012, so the Potts' Corner population is Britain's only hope south of Scotland. And there are fears that climate change may eventually call its number. 'Rising sea levels will become the biggest threat,' Steve sighs. 'Belted Beauty is a moth that lives right on the edge.'

In the wee hours of Good Friday, I collect Wingman Will and Justin Farthing from their Norwich homes. With a long drive ahead – to the eastern Scottish Highlands – we have no desire to crawl like a flightless female moth through bank-holiday traffic. Following an initial career in conservation, Justin became a GP. He wears a delightfully bewildered expression, as if surprised to be alive. Justin and his wife Annie introduced themselves at a school parents' event after noticing me sporting a t-shirt that sulkily claimed I liked 'moths and perhaps three people'. Six months on, it is time for our friendship to stretch beyond the school gate. As cruise-control steadies us north, Justin and I discover a shared heritage of wildlife expeditions to remote parts of South America and associated common acquaintances.

Today's trip has been unexpectedly advanced by ten days. We planned to visit at the very end of April, but a warm February has accelerated Scotland's spring and now the principal object of our trip – the magnificent Kentish Glory – is already airborne. Changing timing, however, consigns each of us to the doghouse. Sharon and Maya fume that I am absent for Easter, while Justin has curtailed his family break and Will persuaded work colleagues to swap shifts. It had better be worth it.

The early signs are good: Scots will enjoy an Easter of untrammelled sun in a vault of cerulean. Then, two hours shy of Aviemore, I receive a message from Speyside naturalist Peter Stronach. To our delight, Peter has successfully located our first target, Rannoch Brindled Beauty, which occurs in only six other European countries. A female is currently clinging to a fencepost a hundred metres off the A9 near Newtonmore. Our car bounces with jubilation as we cut through the miles.

We bypass Perthshire's Rannoch Moor, a famous mothing site: Victorian entomologists used to explore from its lonely railway stop. Rannoch has loaned its name to three moths that, within Britain, are exclusively Scottish. Although too late for Rannoch Sprawler and too early for Rannoch Looper, we are on the money for

Rannoch Brindled Beauty, which frequents boggy ground around woodland in isolated spots from Stirlingshire north to Sutherland. This is another funny moth. Closely related to Belted Beauty, the female is similarly wingless. Pupae can hunker underground for a remarkable four years, scheduling their emergence for suitable conditions.

Once parked, we fringe a pine-dominated woodlot between two bogs then ascend to drier terrain and a line of fenceposts stitching grape-red acid moorland. Peter's instructions are precise, so we count the posts as they stagger away from the trees. Clinging to the thirteenth upright, as specified, is a female Rannoch Brindled Beauty. It resembles a sultana that has acquired weeks' worth of belly-button fluff. In *Enjoying Moths*, Roy Leverton affectionately chides this species for being 'little more than a sporran with eggs'.

We watch her unhurriedly squeeze out single eggs, which stick to the heavily creviced, lichen-smothered wood. Her black, grey-haired body is lined and flecked with sizzling orange, unexpected vibrancy on a creature otherwise designed to evade detection. Intriguingly, the male – not that we see one, despite a return visit two days later – also burns with embers. Leverton suggests the colouration might be aposematic – warning to predators that the creature is toxic and thus best avoided. A Meadow Pipit alights on a nearby fencepost. A flightless female moth would provide an easy snack, yet clearly something is making her feel secure. Perhaps Leverton is right. Either way she furnishes a great start to our Scottish jaunt.

Returning to the car, the A9 ushers us north into Cairngorms National Park. Reaching Granish Moor, we stroll across hummocky grassland where Orange Underwing moths flicker upwards, disguising themselves as last season's burnished leaves at the top of Silver Birches. We follow an open gravelly area along which sun-loving Green Tiger Beetles scurry, before tiptoeing across a railway line, pursued by male Emperor Moths that have caught a whiff of the pheromone lure in my bag. Bouncing over the plum brittleness of springy heather, we search for bearberry and soon spy clumps of this prostrate shrub with green teardrops for leaves and rose-tinted clusters of urn-shaped flowers. This is the sole foodplant of Netted Mountain Moth, a scarce day-flyer. To see it, Peter has advised, visit when the sun shines.

Success comes easily. We swiftly count ten of the skittish creatures amid the bearberry. They are subtle, fragile-looking moths, lead-grey to the naked eye but, through binoculars, sooty with white etching. Their darkness is an adaptation, common among high-latitude and high-altitude moths, to absorb the sun's radiation. This moth is unambiguously solar powered, continuously adjusting position to keep wings perpendicular to the sun and resorting to travelling on foot during cloudy or windy conditions. The adults' spring flight is a risk – Highland weather is rarely predictable – but is timed to optimise caterpillars' start in life, the larvae feeding on the freshest bearberry shoots. Formerly listed in the national Red Data Book of creatures threatened with extirpation, recent surveys in remote parts have revealed sufficient new colonies for Netted Mountain Moth's situation to be deemed less parlous than feared. Nevertheless, it is an auspicious step along our quest for glory.

Our trip north coincides with a worrisome announcement about Scottish moths. Crunching a dataset of five million records, many furnished by thousands of eager volunteers, Butterfly Conservation's Emily Dennis calculates that the abundance of Scotland's moths has halved in the quarter-century to 2014. Ten times more species are 'significantly decreasing' than 'significantly increasing'. Plummeting populations are thought to be the result of habitat degradation, changes in land management and warmer, wetter winters. Dennis softens the chastening with case studies on five moths for which conservation action is underway. This quintet includes Britain's most arresting moth, Kentish Glory – a resonant headliner for our visit.

This large, evolutionarily unique creature – the sole member of the genus *Endromis* – remains Kentish in name alone. Following discovery in England's southeasternmost county in 1741, it proved common west to Somerset and north to Northamptonshire. On a single day in 1858, a hundred were counted at Tilgate Forest, Sussex. Scant years later, England's population was waning. The homogenisation of woodland accelerated the moth's English decline until eventual extinction around 1970. Hope swung to the Scottish Highlands, where the moth had been discovered during the 1880s. The region remains its sole British ramparts.

Fortunately, help is at hand for this nationally near threatened species. In an area synonymous with large, charismatic animals such as Golden Eagle and Scottish Wildcat, conservationists are also taking care of little things that matter. At the time of our visit, Gabrielle Flinn runs Rare Invertebrates in the Cairngorms, a minibeast-centric partnership between several environmental bodies, funded by the European Union. The idea, says Gabrielle, 'is to raise awareness about invertebrates in a unique region characterised by five of Britain's tallest mountains and its largest remaining Caledonian pine forest.' She is overseeing volunteer-run surveys and conservation of six rare insects, including Kentish Glory, which has a paltry three-week flight period in April and early May.

To better understand Kentish Glory, the project has partnered with the aptly Kent-based Canterbury Christ Church University to develop a modern twist on a centuries-old strategy. Entomologists used to place a wild-caught or captive-bred female Kentish Glory in an 'assembling cage' and wait for her to 'call' out her availability to libidinal males. Canterbury scientists have now artificially created pheromone lures that mimic the female's sex scent. As with Emperor Moths in my garden, suspending a pheromone-enriched vial in suitable habitat should encourage any male Kentish Glory nearby to investigate – thus enabling their presence to be recorded. Early results have revolutionised conservationists' appreciation of where Kentish Glory resides. One-quarter of trials in 2017 were successful, with the moth rediscovered in Perthshire after a seventeen-year absence and found at many new and former sites. The following spring was even more successful, Gabrielle says, with 'forty volunteers discovering Kentish Glory in three new ten-kilometre squares and an astonishing seventy-one new one-kilometre squares'.

By chance, one volunteer is a friend's aunt, Mary Laing. Retiring to her family home near Muir of

Dinnet National Nature Reserve in the eastern Highlands, Mary took up mothing with a professional musician's attentiveness to detail. 'In spring 2015, I had the amazing pleasure of catching a Kentish Glory in my own trap,' she recalls. 'I had never heard of it before. Realising how scarce it was made seeing it even more of a privilege.' She became hooked. In 2018, she was Gabrielle's star surveyor, finding Kentish Glory in twelve new one-kilometre squares. 'The hope is that knowing exactly where the moths are will help conservation of their habitat.'

The information gathered enables Gabrielle, the Cairngorms National Park Authority and others to advise landowners about managing properties to benefit the moth. 'We know that Kentish Glory lays eggs only on young birch below two metres in height,' Gabrielle says, 'so we need to restore dynamic ecosystems with regenerating woodland of different ages.' Given that female moths do not disperse far from their birthplace, Gabrielle stresses the importance of connecting fragments of habitat. Fortunately, the UK's largest landscape-restoration initiative, Cairngorms Connect, is doing precisely that. The new initiative has a 200-year vision to strategically manage 60,000 hectares, returning natural ecological processes to forest, peatland and flood-plain. This suggests our desire to see Kentish Glory can generate meaning beyond simple self-interest, and accordingly Gabrielle loans us a lure so that we can participate in surveys. We also make arrangements to run nocturnal moth traps at Muir of Dinnet. Our days will be full, as will our nights, but that's fine: there is no spring-flying moth I yearn to see more.

We start by luring at Granish Moor, trying a number of locations with birch saplings, including paths where friends recently experienced success. Conditions are relatively promising. We are approaching 2–3 pm, the peak flight period revealed by 2017 surveys. The sky alternates between light cloud and sharp sun. Although the air is cool at 8°C, the Kentish Glory's furry thorax (*Endromis* means 'cloak worn after exercise') suggests it to be undeterred by low temperatures. Indeed, nearly one-third of 2017 records were in temperatures below 11°C. A brisk southerly wind should help disperse the pheromones. Whatever, we shouldn't need to wait interminably: two-thirds of records have come within five minutes of a lure being exposed.

We dangle a lure, wait and watch. Nothing happens.

After three minutes, Justin gets self-conscious. 'What are we doing?' he asks rhetorically. 'We're mad.'

After four minutes, I agree. 'Once we reach five minutes, let's try elsewhere,' I suggest.

After four minutes and fifty-five seconds, just as I feint to dismantle vial from branch, a winged ember tumbles an approach, whisking the air a metre out from the lure. My limbs lose co-ordination in the Kentish Glory's presence and I flounce my sweep net – a socially acceptable tool for catching moths, if no longer for butterflies – ineptly hither and thither. A second male arrives. Now Will gets a flap on, waggling his net with similar incompetence. We swivel and swipe in a dance of the deranged, as Justin collapses with mirth, tears streaming. Fortunately, the moths are too sexed-up to flee, granting us time to co-ordinate hand and eye. We eventually enmesh both individuals, then grant them time and space to calm in cooled pots, as we prance around like idiots – a madness of moth-ers intoxicated by success. After ten minutes, curiosity outweighs elation. We sneak a peek at our temporary captives.

To name an animal 'Glory' is to set the bar of expectation high. Yet this lepidopteran royal effortlessly vaults over the threshold. At the front, two harps for antennae pay testament to the vitality of detecting female pheromones. Behind its head, the ermine mantle of a high court judge luxuriates, its initial ring of ivory orbited by a golden throw. A body of russet fur is largely cloaked by inconceivable wings whose wonderment spreads way beyond the demands of function or camouflage. Atop a horse-chestnut canvas, snow flurries and flakes, a black marker pen judders in a toddler's hand, and the odd phoenix bursts into flame. Simultaneously majestic and magisterial, Kentish Glory is justifiably Britain's most widely craved moth.

We test Gabrielle's lure in other promising spots, where young birch needles pin-cushion the heather. The moths might come in from anywhere; a male Kentish Glory has been watched approaching a lure from at least 350 metres away. Two further male Kentish Glory investigate our pseudo-female before it is time to leave.

We relocate two hours east to Muir of Dinnet, running traps throughout three nights. Joined by fellow Norwich moth-ers Christine Steen and Dave Holman, we lug my new generator and four traps from the road to a spot between birch woodland and heathery moorland. Besides a trial run in coastal Essex three days previously, when Will and I helped survey the scarce but visually unprepossessing Sloe Carpet, this is our first night mothing in the wilds, away from the convenience of wall-socket electricity. Hooking up actinic and MV traps to a noisy, petrol-powered machine, we have little clue what we are doing.

But it seems to work. Checking the traps before retiring, a clattering inside my portable actinic draws attention to two whopping female moths: an argentine Emperor and a sandy Kentish Glory. The power and the glory, side by side. Both are so active that potting them would be cruel and possibly dangerous. We agree not to intervene, instead letting the moths decide whether to settle on the egg-trays or escape into the ink.

After six hours' fitful sleep in the local equivalent of Fawlty Towers, we return, entering the quarter-light of dawn with breath bated. Will either female have stayed overnight? As displaying Black Grouse bubble away behind us, I ease off the roof of the trap. Squinting inside, I find six huge red-panda faces smiling a greeting. Unable to believe my eyes, I quickly replace the lid. Surely not... Surely there are not *six* Kentish Glory in the trap, all equably gazing upwards?

Improbable but not impossible, as Keith Kerr would say. We drink in the vision. The female had evidently 'called' for males, five joining her in overnight confinement. She has selected one suitor and remains conjoined with him. Unusually among moths, Kentish Glory mates face-to-face rather than stretching bodies away from one another. The experience strikes a note that would be replayed the following two mornings. In total, we catch twenty-five Kentish Glory – a far headier return on investment than anyone we know. Four are females – each larger than the males – and there's another copulating pair. We admire each and every individual at length, before carefully placing them on birch trunks until their wing-hieroglyphics merge with the shimmering bark. We have smashed even our wildest dreams.

On the journey south, Justin takes over driving, allowing Will and I to snore away the miles. Somewhere in Lancashire, we come to and realise that Justin has missed the intended cross-Pennine turning. As Justin gestures towards a sign for Heysham, I surmise that his navigational error may be intention rather than oversight. Justin wants a crack at Belted Beauty. It seems churlish to deny him such pleasure, so we reorient towards Potts' Corner.

We arrive as a markedly high tide is beating a retreat. Gone are the interminable mudflats; seawater swamps areas where the Morris boys found males eight days previously. In an hour we count fifteen Belted Beauty. Today's sex ratio is skewed towards those flightless, egg-laying sultana-females. Befuddled by such an unexpectedly glorious end to our trip north, Justin stumbles into an unseen channel, wrenching an ankle in the mud. But even the pain can't wrestle delight from his face.

5
Why H is for Hawk-moth too

Norfolk, Dorset, Cornwall and Suffolk
April–June

I am sitting in the kitchen-diner of a smart new-build house in a village somewhere in Norfolk while house-owner Mark Youles sips from a goblet of blood-red wine and a black moth the width of my face bats around the airspace. In the corner, where you might expect a dresser with for-best crockery, a black-mesh contraption – a cross between a fisherman's basket and a mosquito net for use in the underworld – shuffles portentously. Inside, a sisterhood of Death's-head Hawk-moths is awakening, ready to incite lepidopterror.

Thanks to the 1991 film *The Silence of the Lambs*, which featured a closely related and similar-looking species, the immense Death's-head is the world's most readily recognised moth, featuring a thorax tattooed with a piratic 'skull'. (Even so, the creature didn't look quite malign enough for the film's marketeers, and on the famous cinema poster they replaced the thoracic skull with a superimposed section of Philippe Halsman's photograph '*In Voluptas Mors*', which portrays artist Salvador Dalí alongside a *tableau vivant* of seven nude women forming a grotesque, giant skull.)

An academic researching the role of visual and olfactory communication in butterfly mating, Mark breeds moths at home. His kids love the insects, prompting Mark to do a 'show and tell' at their school. Mark's wife, however, particularly despises Death's-heads, deriding them as 'mice with wings'. Extending an arm into the mesh enclosure, Mark extracts a Death's-head. Clad in luxuriant velvet robes that are riddled with chestnut, she is glorious but undeniably sinister. If this moth didn't already exist, surrealist film-maker David Lynch would have invented it. The skull pattern swells up from her thorax, superimposed like icing on a cake. Eyes of juniper berries bulge over two-thirds of her face, burning red in candlelight. She seems as much mammal as moth.

Enmity and dread of Death's-head Hawk-moths have a history stretching back to the early 1800s and spanning Europe. In France, the bedevilled creature was considered an agent of pestilence that could also blind a man with mere wing-dust. In Hungary, its presence in a home signified imminent death. Very unusually among moths, it issues sounds: when feeling threatened, it squeaks through its proboscis – a straw-like tongue – rapidly inflating and deflating air like an accordion. In Germany, this was deemed an audible warning from Death itself.

Such unwarranted associations have sustained the moth's etymology. Vernacular names across Europe are based around the death's-head adornment. The scientific name *Acherontia atropos* exhales diabolical connotations. In Greek mythology, Acheron is the river of pain across which Charon ferries souls of the dead into the underworld. One of the three witch-like Fates, Atropos ushers death by snipping the thread of life.

Souls are pertinent here. Death's-head Hawk-moth has helped engender widespread cultural attachment between moths and the human spirit. Notably, souls of the dead have been thought to escape from the earthly body in the form of moths – a reference to the insect's metamorphosis and subsequent 'escape' from a sarcophagus-like pupal case. In his 1819 poem 'Ode on Melancholy', John Keats refers to the 'death-moth' in terms of being a 'mournful Psyche'. In *The Silence of the Lambs*, the calling card of serial killer 'Buffalo Bill' is ostensibly an *Acherontia* pupa (although actually those of a different hawk-moth, Tobacco Hornworm, were used), deposited in the mouth of the murder victim. The symbolism – of Bill's desire for gender-related transformation – is morbidly clear.

Back in Norfolk, Mark nourishes the female *atropos*, coaxing her proboscis into a bottle-top of sugared water. Immersing face in liquid, she slurps greedily. Unlike the preposterously long 'tongue' of most hawk-moths, the proboscis of a Death's-head is short and sturdy. This suits its purpose of puncturing solid beeswax to reach honey within. These taste buds provide an evolutionary explanation for other features that strengthen its prospects of successful hive-raiding. The 'skull' pattern of what eighteenth-century entomologist and author Moses Harris knew as the 'bee tiger' is said to mimic a queen bee's face. A yellow-banded abdomen further reassures sceptical guard bees that this colossal visitor is benign – one of their kind. Scent may strengthen the hoax, as the moth mimics the smell of fatty acids in bee skin, thus rendering it unthreatening. Thievery is actually as nefarious as Death's-head reality gets.

Mark's moth raises her wings, flashing sulphur undergarments. That people breed moths is one of this year's revelations. I learn that some moth-ers favour such practices, harvesting eggs or caterpillars of wild moths to rear through to adulthood. This isn't my thing; I prefer *wild* in my wildlife. It's also illegal if conducted without permission on a Site of Special Scientific Interest, which many moth locations are. But I appreciate that ex-situ breeding facilitates understanding of a species' life cycle. Mark's approach is very different. Although Death's-head Hawk-moths occasionally breed in Britain – their caterpillars were traditionally discovered nibbling potato plants at harvest time – Mark's stock is of captive origin. He stresses that the practice is entirely legal. Anyone can buy eggs through the internet for a pound per life.

Notwithstanding such accessibility, this particular moth – the ultimate something-of-the-night insect – exudes a whiff of the cult, even the occult. I subsequently establish contact with someone operating under the pseudonym 'Slaughtered Sacrifice'. I explain my interest in learning more about those who breed moths and why. My correspondent retreats into silence as quickly as they appeared. I am denied entry to this particular underworld.

Although their motivations may differ, Death's-head Hawk-moth is as legendary among moth-ers as it is among breeders. It is not an extreme rarity, yet every moth-er yearns to catch one. Mark Tunmore, founding editor of the popular insect magazine *Atropos*, honoured the Death's-head in the publication's name. In its tenth-anniversary issue, he recounted his personal, cross-decade quest to encounter what he termed 'the king of moths', recalling 'a maelstrom of emotions' upon learning that a friend had spotted a Death's-head outside one of Mark's traps set on a Dorset heath and potted it for safe-keeping. Although Mark felt 'perhaps a little excitement' at being so close to his quarry, this was swamped by 'overwhelming horror' that the moth had been constrained, plus 'some anger that [he] had been robbed of the experience of catching [his] own *atropos*'. 'To the astonishment of all present,' Mark declined to see the moth, packed up hastily and departed abruptly. Albeit for understandable reasons – the moth could have flown off – Mark had been denied the exhilaration of personal discovery, and he refused to contaminate the purity of his dream. 'Years passed and *atropos* remained elusive,' he continued. But one September, finally, wrongs were righted. On the wall above Mark's garden trap, 'in the exact place where [he] had long imagined seeing one', a Death's-head Hawk-moth was resting. Here was 'the excitement that had been missing in Dorset; *atropos* had come home to roost'.

My own reverie of catching this underworld legend remains to be realised. The only wild *atropos* I have seen was in repose at Dorset's Portland Bird Observatory, having been trapped by warden Martin Cade. The only Death's-heads that cross my path this year inhabit Mark Youles's home. Now on a sugar high, the female whirrs upwards, chartering the kitchen airspace. She clatters into cupboards, whacks into walls. But after two minutes she goes suspiciously quiet. She has hidden herself somewhere.

'I really need to find that moth,' says Mark, 'or my wife will kill me.'

From Britain's biggest hawk-moth to its smallest. In mid-May, Will and I join Alick Simmons at Powerstock Common reserve in Dorset's undulating north. Alick and I were once colleagues on the environment side of the civil service but left simultaneously, partly out of disgruntlement with the government's attitude to wildlife. Formerly the country's deputy chief veterinarian and perpetually a bearded Scot-by-upbringing, Alick has become a trustee of Dorset Wildlife Trust, which manages Powerstock, a stronghold for the nationally scarce Narrow-bordered Bee Hawk-moth.

Britain harbours two resident species of bee hawk-moth. Both bust the family's nocturnal paradigm by being day-flyers; their genus, *Hemaris*, means 'of the day'. The two bee hawk-moths are also mimics. Indeed, they make trickery into an art form. With their fuzzy plumpness, transparent wings, diurnal buzzing flight and taste for nectar, they are mock-bumblebees, specifically copying the gingery Common Carder Bee.

The point of mimicry is to be noticed. Bee hawk-moths' strategy is to persuade predators that they are harmful animals protected by toxins or weapons – in this case, stinging bees. They are Batesian mimics, named after Henry Walter Bates, a British naturalist who proposed the theory after observing Amazonian butterflies. The approach seems cunning, but there's a catch. Bee hawk-moths must avoid being overly successful. If the mimics outnumber the species they imitate, predators would take the risk of munching. The ruse would be rumbled.

Broad-bordered Bee Hawk-moth is the commoner of Britain's duo – I have already watched one at a Norfolk country park – but it is Narrow-bordered I crave this year, a species I have only previously glimpsed once. Narrow-bordered has declined substantially in recent years, now being almost entirely restricted to pockets of western Britain such as Powerstock. A former ancient common and royal forest, Powerstock is principally managed to benefit its rare insects, notably Marsh Fritillary (a rare butterfly, whose caterpillar shares a foodplant with Narrow-bordered Bee Hawk-moth) and Long-horned Bee.

We seek hints on where to look from Devon-based naturalist Karen Woolley, who has recently visited. 'There are few nectar-rich plants out currently, so it's best to concentrate on patches of Lousewort,' she suggests. 'Try to visit early in the day while they are still fuelling up.' Such advice is worth heeding, but nevertheless it is midday by the time we start ambling west through Powerstock's grassy, woody mosaic, where the blend of habitats bears witness to centuries of changing land use. In the mid-1800s, commoners were deprived of long-held grazing rights, enabling trees to regenerate over pasture. The First World War's demand for wood ridded Powerstock of larger trees, which were shunted south-west along a now-disused railway line. In the 1950s, charcoal-burners cleared high ground before government foresters swaddled the old common with conifers. More recently, Dorset Wildlife Trust assumed control, divesting Powerstock of this environmentally depauperate monoculture and carefully grazing grasslands into flower-rich insect heaven.

Exploring one path, a Common Lizard scuttles towards shelter as Willow Warblers melodise the canopy. After an hour, we bimble into a becalmed, triangular suntrap that screams invertebrates. We surf the lawn's subtle incline, excitedly remarking on the sudden profusion of Lousewort. Hugging cattle-cropped turf, its serrated, garnet leaves tussock below rosaceous petals. This bodes splendidly, and within five minutes Alick yells: 'Got one!' I rush over, but find my friend alone, bereft of hawk-moth. 'It was here just now, I swear… but it simply vanished,' he mumbles, confused. This sounds bang on for a bee hawk-moth. Momentary and illusory. Here today, but gone today too.

Ten minutes later, a tracer straight-lines through my gaze, scudding marginally above Lousewort level. It scours a palinopsic trail through my mind's eye; I know it was there, but it's not there now. Then it returns. Right here, right now, a Narrow-bordered Bee Hawk-moth is nectaring beside my left boot, its proboscis reaching inside a Lousewort. It is the hovering that betrays the imposter's identity; a bumblebee would alight on its nectar source. As I alert the others, it engages reverse gear to back away from the spent flower, then zips off. Its flight is direct, unequivocal, emphatic – divest of the bee's jerkiness.

For an hour we follow two bee hawk-moths around the grassy triangle. One eventually alights, affording prolonged close views. Its fuzzy body is as sandy as Alick's barnet. A solid black belt circumnavigates the midriff and the abdomen tip bursts Dutch-orange. But it's the wings that truly astonish. I gaze through their transparency at the plant beyond, as my eye marker-pens compartments and borders.

We depart a happy trio.

This is a year that delights conservationists too: nationwide, Narrow-bordered Bee hawk-moth is resurging. It regains Cornwall after a decade-long absence, returns to parts of Wales where it has been absent for up to a century, and surprises moth-ers at many new Scottish sites. George Tordoff of Butterfly Conservation Wales primarily ascribes the year's high numbers to 'the result of a good breeding season in the hot summer of 2018'. But it's no coincidence that many new abodes are where his wildlife charity and others have diligently improved habitat for Marsh Fritillary. It may not have been a moth that impelled the work, but everyone is cock-a-hoop that it is a beneficiary.

On the final evening of the Whitsun break, I abscond from familial duties and meet the journalist and author Patrick Barkham and his father in a Norfolk coastal car park in order to search for something rare. Patrick is familiar with such madcap quests: a decade ago, he spent a year criss-crossing Britain to see all our breeding butterflies and related the experience in a bestselling book, *The Butterfly Isles*. Like many before him, Patrick's entomological interests are tentatively expanding into moths, partly encouraged by the generations above and below. His father, John, is a former university professor who traps moths in his Devon garden. Patrick's young daughter Esme, meanwhile, is hooked on hawk-moths, even coaxing her parents into letting her breed what she calls 'Deathies'. If it wasn't the night before school, she would be here too.

Patrick and John emerge from their car, eager moth-ers exiting a co-habited pupa, and we set off, tripping over Common Toads while a Grasshopper Warbler reels out its fishing-line song. Our walk passes quickly amid chat of wildlife and children, individually and in combination. We celebrate moths: 'Even more beautiful, exciting and complex than butterflies,' Patrick confesses.

But we also lament moths: 'If only there weren't so many confusingly similar species,' Patrick groans.

We reach our destination – erratically vegetated sand dunes, scented with sea salt – a good half-hour before dusk. The air is appreciably mild, which is good for moths. But it is also turbulent, which is less so. Fortunately, our quarry – Bedstraw Hawk-moth – is an accomplished aeronaut, so I am cautiously optimistic. A rash of records along the Norfolk coast this week further strengthens hope.

Better known as an immigrant, Bedstraw Hawk-moth is thought to have become resident on parts of England's North Sea coast by 1987. Perhaps milder winters are enabling larvae to overwinter successfully – or perhaps it has lived here longer, simply undetected. Certainly, there are old records of populations temporarily establishing after mass immigrations – at Wellington College, Berkshire, during the 1930s, and at Buxton Heath, near my Norwich home, in the 1950s. Whatever its status, Bedstraw sites (and often sightings) are hushed up – understandably so, given that at least one population was reportedly eliminated in the 1990s after being ransacked for caterpillars. Because of such pressures and its overall rarity, Butterfly Conservation deems the species nationally threatened – 'vulnerable' in official parlance. This renders our visit a gamble. We don't know for sure whether Bedstraw Hawk-moth resides at this clandestine location. Even if it does, we don't know precisely where to look or whether it is even airborne yet. Like the hawk-moth, we're winging it. My excuse of a strategy consists of locating flowering honeysuckle, then waiting and watching. But it works. And how.

No sooner do I heave my rucksack from my shoulders than Patrick's enquiring eyes track a hyper-speed streak of rose and clotted cream. He follows the movement towards blushing honeysuckle entwining a crown of bramble thorns and – wow! – there indeed is a Bedstraw Hawk-moth refuelling aerially. As we gawp, another appears. And another. Each is a hovering blur, impossible for the eye to freeze-frame, only its unravelled proboscis in contact with the honeysuckle's floral trumpets. In the gloaming, splays of blossom and Bedstraw fuse in colour and pattern. Their fates are coupled too: the plant requires the moth's pollination service just as the moth needs the plant's nectar.

Father and son appreciate the hawk-moths differently. Patrick is entranced by the visual. 'They're like fighter planes,' he says. 'And I love the way their eyes flint in the light.' John's reaction betrays the ecologist. 'What an amazing creature to have emerged from the sand,' he observes, in reference to the substrate where the moth conducts its final metamorphosis. 'That mode of flying must require a lot of fuel,' John adds. 'Its energy consumption must be close to that of a bumblebee.'

Individually and collectively, we observe, admire and reflect for a full hour. Then bedtime beckons and we return to the car park, light of step through the tenebrous night.

On our Bedstraw jaunt, John Barkham rejoiced in the abundance and accessibility of many hawk-moths. 'It's so pleasing,' he said, 'that hawk-moths are relatively common so you can expect to see them.' When I first picked up a moth field guide, I surmised the opposite: that beasts as fantastic as hawk-moths would never be found by mere mortals. Such wondrous creatures – impressively large, irreverently shaped and irrepressibly coloured – would remain the preserve of the entomological elite. How wrong I was. Since starting mothing, I have seen eight species in each of our two urban gardens. Hawk-mothing starts at home.

Hawk-moth season stretches from mid-May to August. Our inaugural garden hawk of the year is Lime Hawk-moth. The year proves impressive for this art-deco creation; my catch totals half a dozen, triple the annual average. Friends experience a similar boom. Of all Britain's hawk-moths, this is my favourite. Thrusting raggedy wings away from its body, moth morphs into leaf. A broken band mid-wing further disrupts its shape – an optical illusion primed for the dappled light of shrubbery. As its name suggests, Lime Hawk-moth is green, although the moniker actually reflects the caterpillar's foodplant. And because lime trees often line urban boulevards and pimple suburban parks, Lime Hawk-moth has long been regarded a townie. In his 1937 book *A Moth-Hunter's Gossip*, P.B.M. Allan designated Lime Hawk-moth as 'either an aristocrat or a snob' on the grounds that the moth favoured posher parts of town. Allan is right: Lime Hawk-moth is the only moth to have ever graced the patio of my sister's flat in chichi Hampstead.

Eyed Hawk-moth also garners plaudits this year. Norfolk moth-er Keith Kerr gifts Maya four pupae so she can midwife the final, seminal stages of their life cycle. The size and shape of a Hedgehog poo, the aubergine-purple pupal case is armoured like an armadillo shell and textured with the brass-rubbing imprint of future antennae. The pupae are the outcome of eggs that a female laid in Keith's trap the previous spring. Lovingly, Keith raised them through successive sizes of caterpillar, 'replenishing food twice daily to keep up with their growth rate and clearing up their poo'.

After weeks of impatience, Maya's daily spring from bed towards soil-filled ice-cream tub results in an excited tweenie scream. 'Dada! It's hatched! I'm a Mummy!' Glowing lilac with freshness, an Eyed Hawk-moth clings to the stick Maya provided specifically for its pre-flight warm-up. The final stages of transformation are strenuous. The moth must crack open the pupal shell then squeeze its way out. The prison-breaker then tunnels upwards through several centimetres of soil to reach air, before crawling up a vertical surface such as a plant stem. Pumping haemolymph (a blood-like fluid) into the veins of flaccid wings, the moth inflates them like a paddling pool. Only once its wings are dry can the moth risk its maiden voyage.

Before it departs, Maya and I investigate the Eyed Hawk-moth's three-part strategy for staying alive. Crumpled-leaf wings, blemished by pseudo-shadows, suggest it avoids detection through crypsis. Then I gently stroke the moth's thorax, pretending to be a predator that has fathomed its disguise, and in response the moth flashes its hindwings, startling us with a pair of powder-blue, disconcertingly bloodshot eyespots. Maya's instantaneous reaction is on the money: 'If I was a bird, Dada, that would scare me away. I want to eat, not be eaten!' I elaborate that the 'eyes' serve a further purpose: like those on a Peacock butterfly, they direct any persistent predator's attention to a non-vital part of the body. A pecked thorax means game over. But if the moth merely loses part of a hindwing, it can still operate.

Two days later, a second Eyed Hawk-moth emerges. But something has gone badly wrong. This moth is unable to inflate its wings, which remain scrunched crêpe paper. The moth can run and it can hide, but it won't ever fly. Its chances of mating are negligible, its existence thereby futile. The final two pupae don't even make it this far. They remain immutable, each an invertebrate Han Solo frozen in carbonite by Jabba the Hutt. Maya mourns. One out of four seems a sorry success rate.

Poplar Hawk-moths – my 'Ur-moth', the creature that drew me to the flame – are usually regular visitors to our garden. This year, however, they underperform and we see just half a normal year's count. Sitting by our trap one night, I watch one arrive cumbrously as the MV bulb casts its spell. It flumps around the patio like the obese cockchafers that accumulate in May – and quite unlike the rapid, deft flight of other hawk-moths that pinball into the trap with a satisfying thwack. In 1720, Eleazar Albin wrote in *A Natural History of English Insects* that these moths are 'very swift of flight, for which reason they have the name of Hawk Moths'. Had Albin been acquainted with Poplar Hawk-moth, I wonder whether his description might have been more nuanced.

The next morning, Maya extracts perhaps the same Poplar Hawk-moth from the trap. It clings tightly to her finger with tiny clawed Velcro-feet. Maya regards her novelty ring with the wonder she normally reserves for new dolls. Gently, she releases it onto a rose branch. The moth holds its wings out from the body, hindwing protruding in front of the forewing. Ostensibly a desiccated leaf, it stays put all day, unmolested by local blackbirds. When Maya returns from school, she brushes the moth to check it is actually alive. In response it emphatically lifts and plunges its wings, flashing a cinnamon patch on its hindwing by way of warning. Maya leaves the hawk-moth be. An hour after nightfall, it has reclaimed the skies.

Maya and I may be smitten by hawk-moths, but their sheer bigness can alarm the uninitiated. In summer 2018, tabloids ran stories about the awakening (and implied invasion) of 'giant sex-crazed moths' that 'look like something out of a horror film'. The *Daily Mirror*'s online story – which amounted to nothing more than someone spotting a single Poplar Hawk-moth outside a McDonalds and being frightened by its size – was shared by 43,000 people. 'I was a little scared in case it flew at me,' the observer told the press.

Mottephobia – the fear of moths – is no recent phenomenon borne of nature-deficit disorder, although that surely contributes to modern ignorance. In his 1860 book *British Butterflies*, Victorian naturalist William Coleman wrote that he had never met anyone 'weak-minded enough to be afraid of a butterfly', but had observed people 'exhibit symptoms of the greatest terror at the proximity of a large Hawk-moth'. Coleman recounts listening to 'the grave recital … of a murderous onslaught by a Privet Hawk-moth on the neck of a lady, and how it "bit a piece clean out"'. Coleman's efforts to defend the moth, on the basis that it lacked a mouth and therefore could not possibly bite, fell on obstinate ears.

In inspiring contrast – as Patrick Barkham, Mark Youles and I have each found with our daughters – hawk-moths make perfect vehicles for engaging children with nature. Maya's birthday falls in hawk-moth season, coupling opportunity with motive. Regular party invitees are now accustomed to Maya's hawk-moth parade. Some shy away, fearing the unfamiliar. Others exude confidence, lying moths on their palm, even popping them on a ticklish nose.

Privet Hawk-moth usually proves a crowd-pleaser. It is Britain's most immense common moth, so size alone provokes gasps. But the party trick involves a minatory parting of unfeasibly long, oak-bark brown wings to disclose a coral-barred thorax. In truth, this purported warning of toxicity is merely a bluff. Although I don't have one to show, I suggest the tennis-ball green caterpillar is even mightier. Its habit of rearing up like a Sphinx prompted appropriation of the mythical creature's name as an alternative moniker for the hawk-moths.

If the Privet Hawk-moth doesn't work, even the most recalcitrant child thaws when Maya presents her favourite moth. Elephant Hawk-moth is winged bubblegum, and I am yet to meet a six-, seven-, eight- or nine-year-old girl who doesn't fall in love with it. Anyone who dismisses all moths as little brown jobs need look no further to see the light. Preposterously candy-pink and golden-green, the intensity of its colouration puts every

British butterfly to shame. This winged pachyderm was the first moth that author and naturalist Simon Barnes found; he described it as 'looking like a bird of paradise'.

Without a moth trap, you'd never know that Elephant Hawk-moths existed. With one, you enter a colourful new realm. Normally, herds of 'pink elephants' pack our garden trap – up to twenty per night. This year, however, something is awry. Elephants are both fewer and later. One friend suggests the previous summer's heatwave-baked ground may have hampered underground burrowing, thereby delaying pupation.

The Elephant has a bonsai cousin, appropriately named Small Elephant Hawk-moth. The scientific name – *porcellus*, meaning little pig – seems an apt description for the stubby adult moth, but actually derives from the caterpillar's porcine appearance. 'Small Ellie' is our rarest garden hawk-moth, deigning to appear just once a year on average – and this year continues the trend. Such a paltry showing doesn't disappoint, however, as a windy night trapping amid Lincolnshire sand dunes sees an astonishing fifty individuals sweeten our egg-trays. 'You can never have too much pink, Dada,' Maya observes when I tell her the following morning.

This is not something she could say about the final resident hawk-moth frequenting our garden. Visually, Pine Hawk-moth is decidedly uninspiring, its dusty greyness adulterated with duskier smudges and clusters of short black lines. Pop it on a pine trunk, however, and you understand the rationale. The moth blends in remarkably; even those short black lines resemble errant pine needles. Only one enters our garden trap this summer, but I encounter it readily elsewhere, including in anomalous, conifer-free places such as a Norfolk fen, a Kent shingle beach and a Suffolk saltmarsh.

Such ostensibly mysterious records are consistent with the Pine Hawk-moth's British history, which, in 1937, even P.B.M. Allan admitted was distinctly puzzling. In the early nineteenth century, the British population was thought confined to Scotland – a most unexpected region for a fundamentally continental moth. This prompted mutterings about the claims' veracity. In similar vein, there were suspicions that a population of the 'Fir-tree Arrow-tail-moth' found in Suffolk towards the century's end derived from illicit or inadvertent releases. In 2002, detective work by Colin Pratt – a man so intrigued by hawk-moths that he named his house 'Oleander', after the tropical hawk-moth – concluded that immigrants from the Continent probably started many colonies, but that the major expansion since the 1920s originated from Dorset and followed the spread of commercial conifer plantations. So often the bane of biological diversity, needle-rich monocultures have helped one moth, at least. Between 1980 and 2016, the Pine Hawk-moth more than doubled its range. Even so, it remains a long way south of Scotland.

Maya claims to hear our final garden hawk-moth before seeing it. 'Hummingbird!' she cries. Maya heard its wings vibrate – they reputedly beat up to eighty-three times per second – hazing a hum like the moth's feathered namesake. She gesticulates towards our buddleia, the scraggly haired realm of showy nymphalid butterflies such as Red Admiral and Small Tortoiseshell. A rotund grey fur-ball is suspended in mid-air, its proboscis straining forward into a nectar-saturated bloom on the drooping lilac raceme. Still levitating, the moth swings sideways to infiltrate another flower, this time flashing blazing orange hindwings and, to Maya's glee, a black-and-white bum. Then it is abruptly absent, leaving breathless devotees to contemplate its wake.

Hummingbird Hawk-moth is a moth that gives generously, if briefly, on lazy summer afternoons, its fizzing energy compensating for our prostrate lolling. It loves garden borders and sun, but shuns moth-traps and night. The 'hummer' has long been appreciated as a regular immigrant to Britain, P.B.M. Allan referring to it as 'an ambassador among moths' in *A Moth-Hunter's Gossip*, given its cross-Channel flits from France. Such familiarity has never bred complacency. Quite the opposite. Swedish author Fredrik Sjöberg is no moth-er, but in *The Fly Trap* he admits that even if you see this moth just once in your life, you never forget it. Repeated sightings confer unfettered joy – even ecstasy, as espoused by Virginia Woolf in her essay 'Four Figures'. Woolf was a self-confessed 'bug-hunter' who adored observing the moth on childhood seaside holidays in Cornwall's St Ives. In *The Book of a Naturalist* (1919), W.H. Hudson quotes a correspondent whose childhood was similarly blessed by the 'merrylee-dance-a-pole', whose appearance presented the impression of 'glory, brilliance, aloofness, elusiveness', suggesting it might hark from another world.

I revere a dozen 'hummers' this year, including four in our garden and one, by fortuitous coincidence, in sand dunes cosseting Woolf's St Ives Bay. Every encounter is precious, every disappearance mourned.

Mid-June sees the year's first moth twitch – indeed, the furthest I will drive out of my way all year to see a moth that someone else has caught. It comes at the end of a long day already heavy with hawk-moth. Rising at 2.30 am, I help empty a set of traps on Norfolk's coast then another amid its marshland, the latter furnishing seventy-four individual hawk-moths of five species. While enumerating the Broadland Eyeds and Elephants, we receive notice that Suffolk moth-er Matthew Deans has uncovered a Striped Hawk-moth at Bawdsey Hall.

All hawk-moths are exciting, but those that speak of distant climes are arguably even more so. In *A Moth-Hunter's Gossip*, P.B.M. Allan yearns to see what he calls the 'Rayed Hawkmoth'. He describes it as an itinerant voyager with 'as great an itch for travel as a Scotsman', an animal with unbridled wanderlust whose peregrinations transport it even towards south Asia. A desert nomad prone to occasional demographic and distributional explosions, Striped Hawk-moth comes packaged with thrills and spills. After heavy winter rains in arid North Africa, populations sometimes surge such that thousands tsunami northwards into the European spring and summer. In such years of mass emergence, many may reach Britain: 384 in 2006 alone, according to moth-migration buff Sean Clancy. In other years – of which this may be one – it is virtually absent. This is a moth to see the moment an opportunity arises.

Late afternoon, Wingman Will picks me up in his sporty new number, shimmying us southwards across East Anglia's principal border, passport in hand. Any enmity between Norfolk and Suffolk is largely tongue in cheek, of course, which is why Matthew is happy to share his catch with us as well as Cambridgeshire naturalist Mark Hows. We meet in Bawdsey Hall's car park. Matthew's friends David Hermon and Charlotte Baldwin run the place as a boutique bed and breakfast. Half a mile from the waves, the Hall is an eighteenth-century hunting lodge built for the Tollemache, who were Suffolk nobility. David took it over eight years previously, converting a derelict site into something truly splendid, matching plush country-house accommodation with immaculate garden wildlife. Guests can watch from bed as badgers hoover peanuts scattered across the lawn. The TV offers live feeds of Blue Tit families, courtesy of nest-box cameras. There's an open-fronted hide from which photographers can take floodlit photographs of Tawny Owls. And Matthew gladly shows anyone interested the Hall's moths – of which there are plenty, thanks to an extraordinary set-up.

In discrete sections of the garden or nestling between derelict barns and run-down vintage cars, Matthew runs eight moth traps in fixed positions, tapping the Hall's mains electricity via an intricate network of underground cables. The tessellation of buildings, trees and scrub means at least one trap is always sheltered from even the strongest wind. In this area of dark skies, with no competing illumination, the Hall stands out – and moths flood in. Matthew knows he has it good. 'David or Charlotte even switch on the traps each night. All I have to do is come down in the morning, log the moths, release them and head to work,' he says. 'Although I do encourage people to come and stay here, which helps cover the costs of electricity powering the traps.'

We don't overnight but resolve to return. For now, our attention is swamped by the desert wanderer that Matthew has laid reverently on a lichen-rinsed plank. Shaped like an arrowhead, this hawk-moth's long wings sweep back towards its origin. The bulk of the wing tone recalls the brown hood of the misnamed Black-headed Gull. A broad white-wine stripe splits each wing to its apex, dissected by slender white fishbones. The thorax sports ivory go-faster stripes, the longest flowing from the moth's enormous gemstone eye. The hawk-moth briefly parts its wings, hinting at a hindwing of rose underwear, frilled with black lace. This is a supermoth, an epic traveller with film-star looks.

'It's the sixth *livornica* I have caught here across sixteen years mothing at Bawdsey,' Matthew says, using the moth's scientific name, 'but the first for four years.' Brief mastered, Matthew is as generous with detail as he is with his time. He rattles off the dates of each individual. The trap that held this particular moth has an enviable track record for hawk-moths. 'It has caught a dozen species,' he says, a lofty total that he ascribes to its precise location – sheltered whatever the wind direction and with light blocked off from three angles. 'I think moths get blown towards the Hall from the south, are forced up over the building then encounter shelter and bright light.'

Matthew returns the Striped Hawk-moth to its temporary home of cooled Tupperware. He will hang on to the wonder overnight to show friends in the morning, before releasing it to continue its life – wherever that might lead before the Stygian underworld finally calls its number.

6
The Clearwing King... dethroned

Norfolk, Dorset, Nottinghamshire and Kent
May–July

I spend May Day deceiving the duplicitous. Agreeable warmth notwithstanding, it is too early in spring to realistically encounter any of Britain's fifteen types of clearwing moth – creatures that disguise themselves as wasps. But it's also my birthday, so perhaps Norfolk's Buxton Heath will provide a gift. Sharon consents to join me in a spot of 'afternoon dangling' – seeking to persuade a male moth that we are a female of his species.

Like spring's bee hawk-moths, clearwings are Batesian mimics. Britain's clearwings fall into one of three camps. Most have black and yellowish stripes like a solitary wasp. Several are black with a red band, copying spider-hunting wasps. Two species up the ante by mimicking hornets. Their deceptive capabilities extend beyond appearance to behaviour: clearwings hover like wasps and twitch antennae in a decidedly unmoth-like fashion. Little wonder they fascinated novelist Vladimir Nabokov, who lauded them in *Speak, Memory* for having evolved such subtle mimicry that predators stood no chance of seeing through their deception.

This century, finding clearwings has become much easier thanks to the commercial availability of synthetic pheromone lures. In recent decades, pheromones of many insects have been synthesised to help control pest species, in particular. Several clearwings – including, as its name suggests, Currant Clearwing – can infest fruit to the level of crop failure. Affected growers lure males to their death with the artificial scent of a sexed-up female. More amenably, moth-ers may now ascertain the presence of most clearwings by dangling from branches impregnated rubber bungs or plastic vials, pleading science and conservation as justification for temporarily coaxing males into a forlorn libidinal frenzy. To our eyes and noses, there is simply a terracotta-coloured piece of rubber – nothing to disrupt a moth's life. Staring at the bung, waiting for something to happen, is mesmeric – first to the point of being meditative, but eventually to the cusp of madness. Until it works. 'The first time a clearwing lure comes good,' conservationist Ben Lewis told me, 'it's like magic.'

At Buxton Heath, my nominal quarry is Large Red-belted Clearwing, the first clearwing to emerge each year and one of three British species whose range seems to be dwindling. It's a moth nemesis for me. Twice I have glimpsed it, and twice it has evaporated before I could swish a net or focus my camera. The species loves old birch, particularly on dry heathland – as at Buxton Heath. Nobody has ever seen one in Norfolk before 2 May, so an eager individual today would break records. I hang two terracotta bungs ten metres apart among tufty grass pinned with birch saplings teetering above my head and cordoned by waist-high rotting stumps. Sharon watches one lure; I take the other.

Slightly agitatedly, Sharon asks what she should be looking for.

'A chunky black wasp with a red cummerbund,' I call, 'buzzing around the lure.'

'In which case, you better come quickly. There's one here.'

The lure has only been up for thirty seconds! I dash over, narrowly evading a concealed, decomposing bough – birchland's booby trap. Sharon is not fibbing. Norfolk's earliest-ever Large Red-belted Clearwing – and the first recorded anywhere in Britain this year – is interrogating the rubber pretence of a female. The libidinal male can smell her but can't get a visual. I further his confusion by flicking a net, encasing him in mesh and potting him.

His agitation quelled, the moth reveals his gothdom. This clearwing's favourite colour is black, even down to its winkle-pickers. Only cinnamon-red disrupts the uniformity. Flashes on the wing base and the underside of the palps (sensory mouthparts) complement the girth-wrap, which is the primary clue that this moth mimics a Red-belted Sand Wasp. It's a pretty good match, even down to transparent, boldly veined wings of stained glass. It's a wonder of evolution, in my hand.

I liberate the clearwing to reclaim its brittle-heather, silvery-barked domain. The moth that Moses Harris called 'The Bishop' during the eighteenth century is spared further ignominy of getting het up for nothing. I sprint off on a victory lap, fist-pumping and yelling for joy. A Meadow Pipit takes umbrage, flying off. Sharon, however, regards me indulgently. It must be my birthday.

Like Nabokov, I am captivated by clearwings – and have been for three years. When I started mothing, these takes on waspishness were mythical creatures of wonder – unattainable, untenable – mere pictures in a field guide. During May and June 2016, however, I discovered pheromone lures. Initial attempts in intriguing landscapes around Norfolk conjured no sniff of interest. But when I looked afresh at clearwings' ecological needs, I realised those of Red-belted – the slighter cousin of my birthday present – coincided perfectly with two antique apple trees sprawling over our patio from the neighbours' fiefdom. The following day, I hung two lures in the garden, expecting nothing. Within sixty seconds, however, black-and-scarlet wasp-a-likes were buzzing around each. That afternoon, ten Norfolk moth-ers popped over for a cuppa and a clearwing. Red-belted Clearwing was seemingly rare in the county, with four summers scraping a meagre total of five records. Moreover, it was new to everyone, prompting happy grins and energised chat all round. If clearwings lived in *gardens*, that opened up thrilling possibilities. We could make important discoveries without troubling the carbon-counter. Staying at home would become the new going out.

Drunk on moth festivities, I wantonly strung up a lure for another clearwing, Hornet Moth. Again, we expected nothing – particularly as this larger species is active early morning, not late afternoon. But fifteen minutes later, a huge yellow wasp orbited the lure in ever-decreasing circles. Surely not? I swished the net, eliciting a furious buzz from inside the mesh. Some underlying atavistic fear compelled my palms to start sweating. I had nabbed a genuine hornet. Error! But no, this *was* a Hornet Moth, its mimicry so fine-tuned as to encompass sound alongside sight – wings made to whirr like an irate vespid. Delight burgeoned into disbelief-infused ecstasy. Hornet Moths spend their youth inside poplar trees. The nearest poplars were a mile away, so where had it come from? And what else might be enticed to the garden by the wafting of artificial sexiness?

That afternoon, a new cohort of clearwing-surveyors was generated. Over successive summers, we were inspired to deploy lures all over Norfolk, revealing an array of clearwings to be more common and widespread than even dreamt. That day also, a network of moth-ers in and around Norwich was born. Ever since, and now forty strong, we have shared details of our daily catches via WhatsApp. We have served as a mutual support group for moth identification, techniques, locations and more. Together we have tried, learned, discovered. And drunk many a cuppa while admiring moths.

That heady summer, wherever I went, I seemed to encounter clearwings. Some people suggested my clothes had become impregnated with pheromones. (It's not a bad theory; one day a White-barred Clearwing explored my jeans pocket.) Rendered jealous by this success rate, one member of our Norwich Moth-ers Anonymous collective, novelist Ed Parnell, named me 'The Clearwing King'. The moniker stuck.

This year, by late May, I manage four more clearwing species, three at home. Chunkier than Red-belted Clearwing, Yellow-legged Clearwing occupies a more familiar wasp form. Protruding from a body ringed yellow and black are six buttercup limbs. It loves oak, of which a gargantuan example hulks nearby. Our garden includes a bonsai attempt at an allotment, two token currant bushes piercing its vegetative riff-raff. Both are now sufficiently mature to harbour breeding Currant Clearwing, a speck-of-life version of Yellow-legged, and to attract a Red-tipped Clearwing. Being new for the garden, the latter thrills. It rejoices in sweeping scarlet wing-tips, a zebra-striped abdomen and blinding white sides to a splayed 'tail'. Books suggest Red-tipped Clearwing inhabits damp places: riverbanks, carr, ponds and such. What truck it has in Norwich gardens – and I've seen it in several – is unclear. Such mystery is clearwings to a tee: secret lives unfolding in disguise.

Which other wasp-striped enigmas might inhabit my neighbourhood? I try for Orange-tailed Clearwing, but without luck. Its range may have spread to west Norfolk, but it's not here yet. Giggling, I try for Six-belted Clearwing. This is pie/sky territory – a species of chalky or sandy ground, dependent on harnessing sunny blooms of Bird's-foot Trefoil. Not a garden moth. It went unrecorded in Norfolk for 140 years before eventual rediscovery in 2007. Unsurprisingly, I don't succeed. There's no earthly way it can live in Norwich.

Perhaps I should have persevered. Later that week, Kentish Glory partner-in-crime Justin Farthing is necking an off-duty beer in his unequivocally urban Norwich garden, watching the sun bounce off a succession of unrequited clearwing lures. Doubtless as giggly as me, he too tries Six-belted Clearwing. As a sturdy, yellow-and-black-striped insect approaches the terracotta bung, he fears he has imbibed one beer too many. But, no, it *is* a Six-belted Clearwing. WTF?!

Conveniently, Justin lives adjacent to my daughter's school and I'm on pick-up. It would be rude not to... Opening the door, Justin looks even more bemused than usual. He cannot fathom this clearwing's appearance. The Farthings live two miles from the nearest known patch of Bird's-foot Trefoil. There's a golf course three-quarters of a mile away that *might* have the plant, but that still seems too distant. When Joe Burman of Canterbury Christ Church University tested lures for burnet moths, he found their effectiveness dropped off markedly at 180 metres. This clearwing must have journeyed much further. But moths never stop surprising. Ten summers ago two Six-belted Clearwing were spotted in the garden of London's Natural History Museum, presumably having flown miles to get there. It is a truism that we have so much to learn about moths. The duplicitous, in particular, retain much that they have not yet divulged.

Two weeks on, the same species emphasises the same point. Old friend Durwyn Liley, Sharon and I usher our kids to a secluded beach near Lulworth Cove in Dorset. As we labour down a steep path under an unforgiving sun, we bump into wild-haired National Trust ranger Billy Dykes. He grants permission to try luring Six-belted Clearwing among rampant Bird's-Foot Trefoil above the beach – but warns that he's never seen the species here. Nevertheless, within five seconds of Durwyn liberating the lure from his bag, a dozen clearwings are flocking excitedly. Within a minute, there are scores – a swarm of wasp-a-likes at a picnic. None of us has witnessed anything like it. And this for a species that Billy didn't even know was present.

You don't need lures to see clearwings, of course. If you research where to look and season sharp eyes with luck, you stand a chance. Without artificial stimulants, this year I lumber into Currant Clearwing on an allotment, Red-belted Clearwing sheltering on a wall from volleying rain, Red-tipped Clearwing nectaring on bramble and White-barred Clearwing in soggy Alder carr in Norfolk's Broadland. But perhaps the easiest clearwing to see without a helping hand is Hornet Moth.

The Big Daddy of Britain's clearwings makes its larval lair inside a mature poplar tree. If the base of such behemoths is repeatedly hole-punched, leaving cylindrical exit points the girth of your little finger, you're probably in luck. For their clearwings, Victorian collectors relied on spotting such tell-tale circles. Each insect literally eats itself out of house and home, squeezing out of its shelter as a fully-winged adult. One late June morning, following a tip-off, I scrutinise the toe of poplars standing sentry over the River Yare. Above holes on the first tree squats an adult Hornet Moth.

It's been two summers since I last saw the species – probably how long this individual has spent encased in wood – and its size startles. It measures two knuckles' worth of my pinky. It is a fine mimic – a caricature, even, of the stinging vespid. For sure, the moth's true identity is betrayed by lacking the hymenopteran's pinched-in waist and by displaying yellow shoulder pads rather than chestnut. But most predators wouldn't risk close-up examination. Most, but not all. Dame Miriam Rothschild, member of the family banking dynasty and thus the closest that Britain has to a mothing celebrity, once fed adult Hornet Moths to bird predators. The birds tried a nibble then spat the moths out, wiping their bill on bark – a sure sign of something distasteful. This suggests that Hornet Moths – and, by extension, perhaps other clearwings – operate twin-track mimicry. Their first gambit is Batesian ostentation – visually resembling a scary creature. But they back that up with genuine toxicity, communicating honestly that they should not be eaten. This strategy is known as Müllerian mimicry, named after German naturalist Fritz Müller. It doesn't always work; a friend recounts watching Blue Tits waiting beside a Hornet Moth tree and nabbing adult moths the moment they reached the trunk surface.

Nevertheless, the closely related Lunar Hornet Moth presumably adopts similar survival tactics. It resembles its cousin but exhibits a black rather than largely yellow head and a marigold necklace rather than epaulets. This is one of four British clearwings that I have never encountered. Not uncoincidentally, it is also the only British species whose pheromones remain to be synthesised. Seeing it therefore relies either on pure serendipity or on visiting trees known to harbour larvae early on July mornings and hoping to coincide with a freshly emerged adult before it absconds. I ascribe three July dawns to checking willows upriver from the Hornet Moths. Each morning is cobalt and peaceful, but every runnelled, twisted old *Salix* remains resolutely free of Lunar Hornet Moths.

I'm not overly dismayed, however. My sleeve conceals an ace: family Lowen is booked into Dorset's Portland Bird Observatory mid-month. In 2017, Lunar Hornet Moths were discovered around stunted willows in the Observatory garden. How long they had been there and where they had come from are questions to which warden Martin Cade can give no certain answers. Early one morning, Martin kindly escorts me into the garden's inner sanctum – an area otherwise out of bounds. We scrutinise several wizened willows, squat versions of the Yare's giants. Martin's index finger points towards several holes. Below one shudders a volcano of frass – a cone of wood-dust ejected by a Lunar Hornet Moth caterpillar before its adult journey to sunlight. But of the evacuee, we catch no sight. For this particular clearwing I will be obliged to wait another year.

At least I got a crack at Lunar Hornet Moth. I don't get any such opportunity with the similarly *Salix*-loving Sallow Clearwing. Tinier than even Currant Clearwing, this smidgeon of existence follows a strict two-year life cycle, emerging as adults in even-numbered years only, which rules out this year. For what proves to be my sole new clearwing of this Big Moth Year, I must search for something Welsh in England.

As Kentish Glory and New Forest Burnet illustrate, naming a moth after a location risks subsequent anachronism. The fall is potentially even greater when doing so in honour of a country. Granted, Scotland has got away with Scotch Annulet and Scotch Burnet – moths that, within Britain, remain exclusive to that nation. Wales hasn't been so fortunate. Not only is Welsh Wave widespread from south Devon to Sutherland, but Welsh Clearwing occurs in the two other British nations as well as its motherland. Worse still, it was lost to Wales (and to England) for sixty years. As recently as 1988, it was considered extremely rare, persisting only in a few remote Scottish sites. That year, fortunately, it was relocated in Wales and subsequently found over a fair swathe of the country. In 2005, it was rediscovered at Staffordshire's Cannock Chase. Three years later, a second English site was chanced upon: Sherwood Forest in Nottinghamshire. And thus, a week shy of the year's midpoint, Will and I find ourselves beneath the papery boughs of knobbly, thick-set Downy Birch trees in a sunlit, grassy opening of Robin Hood's chancellery. It has just gone 8.30 am – premature for rousing Welsh Clearwings from slumber, we suspect. Indeed, it is probably too early full stop, as nobody has yet encountered the species anywhere in Britain this year. But you gotta be in it to win it. And within twenty minutes, two male Welsh Clearwing are lazing towards our lures.

They are unhurried critters, disdaining the hyperactive buzziness that defines other clearwings. They're large too, enabling us to watch one stealth-bombering in from fully five metres away. We are soon joined by James Harding-Morris, a conservationist working on the Back from the Brink conservation programme. Today is James's thirtieth birthday; he is pumped by the gift we have unwrapped for him. Together we gaze at the Welsh Clearwings loitering on the birch bark, lustful enough to have flown in from elsewhere but sufficiently savvy to have determined something to be awry. Each insect is largely black with pale yellow tramlines lining its thorax, two bands traversing its abdomen and a tangerine shock of a splayed 'tail' tip. In ninety minutes we enumerate at least seven, making us quite the band of merry men.

We're in robust company too. Charles Gregson was a Victorian ship-painter who earned sufficiently handsomely to really indulge his obsession with moths. Gregson was integral to a group of Lancashire collectors who saw themselves as rivals to London's entomological establishment. One-upmanship was part of the game, and Gregson was renowned for pushing the boundaries of the possible. So intent was Gregson on possessing Welsh Clearwing that he once paid for some old birches to be felled and conveyed to him by train in the hope that they contained moth larvae. History does not record whether Gregson's outlandish gambit was rewarded.

My clearwing wish-list has a final member, but a wish it shall have to remain. When discussing potential target species with key conservationists and moth-ers, I was asked to exclude Fiery Clearwing. This is one of the UK's eight strictly protected species; in the UK, it occurs only in Kent. Disturbing it – by using a pheromone lure, say – is illegal. 'Even looking for it without a lure is against the law,' warned one adviser, 'because actively searching risks damaging its habitat. Please leave this moth be.'

Such emphatic strictness took me aback. I didn't hesitate to respect the request, of course. The moth's welfare comes first. But the mandate does prompt contemplation of the rationale for the prohibition. After all, I had received no such blanket ban on looking for other members of the legally protected octet – simply words of caution and requests to seek permission. And it's not as if people don't look for Fiery Clearwing. 'They just keep quiet when they see it,' one Kent moth-er admits. Furthermore, although critically endangered, populations of this flame-winged beauty are growing at several sites and it stands to benefit further from Butterfly Conservation's 'Kent Magnificent Moths' initiative. So what justifies the austere treatment?

Collecting might be a factor, as with New Forest Burnet, another of the protected eight. In the early 2000s, Fiery Clearwing foodplants were dug up at a protected site – presumably with the caterpillars inside. Even burgeoning populations can be fragile; one bad summer or wildlife-scorning infrastructure development can flip today's success into tomorrow's failure. But the fundamental difference in treatment, I surmise, must relate to the use of pheromone lures. These have no doubt revolutionised our understanding of clearwing status and distribution, which is critical for efficient conservation. For ten of the thirteen species for which it makes sense to count such things, the number of ten-kilometre squares harbouring the species has swollen since lures have been available. For eight species, this crude distributional measurement has doubled or tripled since 1970.

Purely intuitively, however, luring seems bad for individual moths. Males are conned into expending valuable energy to investigate a non-existent mating opportunity – possibly at the expense of missing out on a genuine female elsewhere. For most clearwings around my garden, this might not matter overly; they probably don't travel further than fifty metres. But the investment would have been greater for my initial garden Hornet Moth or Justin Farthing's errant Six-belted Clearwing.

Ethically, this realisation increasingly troubles me, forcing refinement of what seems an invasive practice. This year, I use pheromones sparingly. As soon as the lure has weaved its magic – and it really does feel like sorcery – I whip it down. In my garden, I operate cautiously for the resident species, seeking to understand the length of their flight period – using lures once a fortnight for just a few minutes. Any longer would be gratuitous and invasive. Later in the year, I am pleased to see Butterfly Conservation issue guidelines on pheromone use, designed to enable the collection of valuable data while minimising disturbance. In this context, prohibiting the use of lures with Fiery Clearwing is understandable – and morally right. If dozens of mothers lured successively at a well-known site, the chances of population-level impact there may be very real. Nobody would want that.

Ethics and intuition are all very well, but what does science tell us? Joe Burman studies how insect science can help solve problems in agriculture and biodiversity conservation. Nobody knows better the implications of using pheromones to monitor rare species, and his experiments have confounded expectations. For Six-spot Burnet, at least, Joe and his colleague David Thackery discovered that pheromones have no negative effect on males and, indeed, may actually help those that have mated live longer, extending their ability to procreate. Moreover, pheromones have no effect at all on how many eggs a female lays. Joe suggests these results 'go some way to demonstrating [pheromones'] safe use on fragile species'. His colleague Ashen Oleander found that having been exposed to synthetic pheromones did not affect how long it took for male Six-spot Burnets to find females or what proportion of males located females. Moreover, when lures were used, males spent *more* time with live females, not less.

It would be presumptuous to extrapolate that Fiery Clearwing would react in the same way as Six-spot Burnet. In other species of moth, experiments have shown that pheromones disrupt mating behaviour, interfere with males' awareness of predators and reduce lifespan. Such issues, say Ashen and Joe, are worth investigating before declaring pheromone monitoring of rare insects to be completely safe and devoid of impact on populations. For now, 'better safe than sorry' seems an appropriate approach with Fiery Clearwing.

This particular moth may be out of bounds, but it would be remiss not to experience something of Kent's clearwings. Will and I make for a garden near Ruckinge, on the fringe of both Hamstreet Woods and Romney Marsh. The Boothroyds moved here twenty-three years previously, soon planting up their 1.5-acre plot that today oozes mature trees, flowers and fruit. In this moth haven, Bernard Boothroyd – tall, gentle and bearded – has identified a remarkable nine hundred species, including an unprecedented nine clearwings. No other British site comes close to this, let alone a garden.

We call in on Bernard three times during the summer. In early July, he presents us with an agreeable surprise: an Orange-tailed Clearwing lured shortly before our arrival. It resembles a large Currant Clearwing, but with an orange 'tail' fan instead of yellow shoulder stripes. As we photograph it in his conservatory, the clearwing suddenly levitates vertically then makes for the expansive window.

It alights by a hitherto-unspotted Yellow-legged Clearwing: Nabokov's beloved wasp-mimics thrive among the Boothroyds' botanical riches. My quintet of garden clearwings now seems paltry. Bernard, not me, is Britain's undisputed clearwing king. I have been dethroned.

***Postscript**: In late June 2020, a trial pheromone lure for Lunar Hornet Moth was made commercially available. This proved a game-changer. A formerly impossible moth immediately became eminently accessible. Following just twenty-five records of adults in Norfolk during the first twenty years of this century, the lures generated records from thirty new sites within a week, swiftly clarifying the species' distribution. I even caught one in my garden. But then, so did Bernard Boothroyd.*

7
If small is beautiful, how gorgeous is tiny?

Dorset, Norfolk, Monmouthshire and Kent
May–July

'Don't be sizeist!'

Step forward Will Soar, passionate defender of all creatures great and small – but particularly small. Will was berating my disdain for the 1,630-odd species of what nineteenth-century entomologists first characterised as 'Microlepidoptera' – size-challenged insects that include two of every three British moths. Will's admonishment came in summer 2016 when we were perusing my garden trap and I dismissively passed him egg-trays dusted with tiny, irrelevant specks of life. 'Micros' weren't my thing. They were too small to see, too difficult to tell apart and seemed to have exclusively Latin names that were longer than the actual moths. As far as I was concerned, they were all Will's.

I was hardly alone in looking askance at micromoths. Many experienced moth-ers intentionally overlook, ignore or malign them. Until the 2012 publication of the first user-friendly field guide to mini-moths, micros were largely neglected by the hobbyist, remaining trapped in the province of specialists. E.B. Ford covered only 'macros' in his 1967 tome *Moths*. Tongue in cheek, he considered this 'an entirely unjustified restriction which will be welcomed by the majority of naturalists'. Ford was right on both counts. Micromoths take effort. 'I don't do micros. Life is too short,' one keen moth-er told me. 'Too daunting to start,' admitted another. In *A Moth-Hunter's Gossip*, P.B.M. Allan ascribed his disinterest in smaller moths as being down to laziness.

But moths are moths are moths. The allocation of moth families to either Micro- or Macrolepidoptera is, as Ford admitted, both artificial and arbitrary. Granted, as their name suggests, macros tend to be larger than micros. But some are as small as many micros, and some well-known moth families treated as macros (such as burnets, clearwings and even the humungous Goat Moth) are actually supersized micros.

The first chink in my permafrost came that September when I caught a little citrus-loving creature from tropical Africa called *Thaumatotibia leucotreta* (False Codling Moth). This was the first ever found in Norfolk, although it had probably stowed away with a fruit shipment rather than reached my garden under wing-power. Nevertheless, discovering a county first without even leaving home hinted at an intriguing return on investment for engaging with micros.

My frost thawed further the following summer as I determined that fun-sized moths could be, well, fun. A friend showed me a truncated moth to which he ascribed a far longer name: *Dyseriocrania subpurpurella*. It looked to be wearing enormous black shades, boasted a mass of straggly black hair and was clad in a heliotrope-dotted, glam-rock golden cloak. If ever a moth recalled Danny, the unsettling stoner from cult film *Withnail and*

I, this was it. Later that day, I gasped at swarms of pop-eyed male *Adela reaumurella* (sometimes called Green Long-horn) dancing in a sunny arena at head height, all yo-yoing while waggling preposterous antennae four times their body length. This boyish dance-off was all about competing – lekking, in ecological parlance – for females.

I became bewildered by the utterly unmoth-like appearance of plume moths. With legs held in an 'X' and slender bodies from which feathery wings stuck out at right angles, these twiggy creatures fluttered as feebly as a gnat. The plumes' take on evolutionary adaptation got stranger still. I was amazed to learn that the caterpillar of the tiny, gnat-like Sundew Plume turned tables on the eponymous carnivorous plant. Subsequent research revealed that the caterpillars avoid getting snared by licking off the plant's sticky mucus, then ate the plant's glandular hairs. This insect eats plants that eat insects. The revenge must taste sweet.

I mourned the all-too-brief life of Water Veneer, hundreds of whose corpses littered the base of my trap. As its scientific name suggests, a Water Veneer's adult existence lasts but a few (hectic) hours. These are the mayflies of the moth world – scraps of life for which midnight tolls once only. One balmy night I watched a silent rave of a thousand males flickering above a pond surface. Later I ascertained that this moth is truly amphibious, its caterpillars and the wingless form of females living happily underwater. Water Veneers are moth madness.

I came to appreciate that garden micromoths could be lookers; they just needed close examination. *Lozotaeniodes formosana* proved aptly named – *formosa* being the Latin for beautiful – given the moth's salmon- and strawberry-pink oscillations. *Phtheochroa rugosana* challenged my preconceptions, being long-snouted, rippled with ruffs of scales and exhibiting a dozen colours. The long, iridescent wings of *Roeslerstammia erxlebella* shimmered bronze, purple and green as the light danced upon them.

Will was right: these micros really had something. So as my year clunked into gear, I made two resolutions. I would treat all moths as equals; micros should rightly feature in my quest. And I would handpick a quartet of glorious, must-see micros that leap out of Phil Sterling and Mark Parsons' *Field Guide to the Micromoths of Great Britain and Ireland*, then make concerted efforts to track them down.

The first target was called Geoff.

As I watched, Geoff fluttered slowly but inexorably away, beyond the thorny crossfire of an impenetrable bramble sprawl. Geoff's proper name is *Alabonia geoffrella*, but many moth-ers revere him (and her) as Geoff. *Alabonia* is an undisputed supermodel on the micromoth catwalk. Yet being relatively common and widespread across England and Wales, this glitterball of neon, gold and bronze feels attainable. Geoff is moth *haute couture*, but available on the high street.

Geoff normally emerges into hedgerows, woodland edge and scrub during late May. It is a morning creature powered by the first rays of sun. Its caterpillars love the dead stems of bramble and Blackthorn. My sole previous meeting with Geoff was a potted individual literally on its last legs, having been over-vigorously swept from the undergrowth. It barely counted. Impatiently, I start looking for Geoff the moment May passes halfway. Doing so involves thinking differently about life. To a micromoth, a single bush can be an island of habitat. Spotting micromoths involves slowing down, constricting your world, focusing so intently on nothingness that passers-by assume you are away with the fairies.

My first seven attempts are unsuccessful. Each failure breeds frustration, erodes faith. Then comes the news that friends have spotted Geoff in Herefordshire and Norfolk. Its flight season is underway. Game on. That night, I happen to be sleeping out in a Dorset woodland, a hub from which luxuriant hedgerows spoke outwards. Such habitat seems plausible for Geoff. As I huddle in the chrysalis of a bivvy bag, neon, gold and bronze glitter through my dreams. An hour after dawn, in a shard of light perforating the canopy and granting life to shrubbery, I glimpse floating gold dust: Geoff.

The moth's shape is unexpectedly mosquito-like – long trailing legs are balanced by extended antennae and modified mouthparts called palps that strain forwards – but there's no mistaking *Alabonia geoffrella* as it skitters through the air over the bramble, out of sight and out of my life. Anguished, I flail the net hopelessly in Geoff's wake before collapsing to my knees, spiky vegetation piercing my trousers and dignity. Wingman Will also collapses, but in hysterics. He may have missed the moth but he's lapped up the year's funniest sight yet.

We need not have worried. We soon find another dozen Geoffs, eight dancing together above a shady bramble tangle. Their flickering lek mesmerises. As I stare, silent attention fixed, an early morning dog-walker asks whether I am OK. My breathless, burbled response ('micromoths… Geoff… he is real, after all…') confirms her suspicions; she hurries off. Lost in wonder, I care not. Each individual Geoff is a bonsai bijou. A golden front half is strip-lit with an electric-blue neural network, before intensifying into a copper rear emboldened with silver triangles and lined with mascara. Even Geoff's shape is remarkable. Long, thick black palps thrust upwards then periscope vertically from an elbow-like joint into miniature white tusks. Behind sighing antennae, a tiny head with minute eyes broadens quickly to fulsome, generously curved hindwing tips.

With my eye now in, I encounter Geoff everywhere – even without searching. I don't look but I still see. Geoff demands my attention in a dark Kent woodland, at a random lay-by on a Norfolk dual carriageway, and along a pavement-flanking hedge on the outskirts of Norwich. I watch Geoff flying at all times of the day, not just early morning. Geoff is the micromoth I have been waiting for.

The Wingman and I up the ante with our second target moth. It will be a considerable challenge. Posing on the same field-guide page as Geoff is a close relative, the sumptuous *Oecophora bractella*. Whereas Sterling and Parsons' mini-moth bible describes Geoff as 'common', *bractella* is 'rare'. Butterfly Conservation's 'Moths Count' website suggests that Geoff is known from fifty-eight of the 112 vice-counties (a geographical unit into which Britain is split for wildlife-recording purposes), but *bractella* has appeared in just twelve and there are far fewer regular sites. I managed Geoff at my eighth attempt but have just one crack at *bractella*.

One late-June night, we place all our chips on the extensive, ancient woodland of the Lower Wye Valley north of Chepstow. Will and I have five traps, but raise chances by joining forces with three experienced moth-ers: George Tordoff of Butterfly Conservation Wales, Will's mate Lee Gregory and my school pal Matthew Hobbs. George is particularly excited about the opportunity to catch *bractella*, a moth for which he has repeatedly tried, crashed and burned.

'For whatever reason, *bractella* seems to need large areas of habitat.' George speaks precisely and urgently. 'It only seems to occur in big old woods, particularly those with dead trees.' The place we have chosen fits the bill, its rocky rise from the River Wye clothed in uninterrupted forest. Its trees are diverse and mature; broad-trunked oaks, Beech and Small-leaved Lime make peaceable neighbours. Woodland has swathed this land continuously since at least 1600. The Wye became a tourist destination as famous as Snowdonia in the early eighteenth century; 150 years later, naturalists learned its wildlife significance. Even so, the Wye's star macromoth, Scarce Hook-tip, remained undiscovered here until 1961. This is now the only place in Britain where it still lives.

We have timed our visit to try for both rarities. The Wye canopy is thick, the pre-dusk light aquatic; the night then proves warm and moist, its darkness compressing. Our traps heave with wings. I blow moths out of my eyes and nostrils; they cloud upwards, dissipating into peaty leafiness. Nevertheless, it takes until the half-hour before midnight for our lights to entice fourteen Scarce Hook-tips down to ground level from their comfort blanket of the lime-tree canopy. This moth's charm stems principally from its shape. Backswept boomerang wings (the 'hook-tip') half the length of a finger recall a small Atlas Moth (an outlandish Asian behemoth). A complementary understated beauty draws from a fruit-machine splay of golden coins mid-wing and purplish black-eyes towards the hook.

Within two hours, our three guests have departed; beds briefly calling their number before work starts. Will and I press through exhaustion for another hour, raising our catch to 115 species. No *bractella* appears, but nor does despondency… yet. The moth flies from the final hour of darkness until early morning, so there is still time. Nevertheless, we feel obliged to retire for the night's remaining slither. I ground a bivvy bag beneath a generous lime as the sky empties its bladder. After what seems like minutes, we struggle into consciousness before slipping downslope into the monsoon dawn. A thousand moths cram rain-lacquered traps, the first four of which are sadly devoid of *bractella*. Despondency no longer feels premature. The final trap, however, yields not just one *bractella* but a gang of four. Success prompts Will into a jig of joy; I sing in the rain.

Our quarry is smaller than Geoff but similarly shaped, even down to its extended, curving palps. *Bractella*'s colour pattern is simpler. It is essentially a moth of two halves: marigold-yellow in front and coal-black behind.

The latter is dusted with blue sparkles and sees a yellow sun rise from the wing edge. It is sublime, a summer cheer to offset an indefatigable storm. Gently, we station the exquisite quartet on the trunk of the same tree that afforded me overnight shelter. They sit tight as we depart.

For creatures so small, micromoths court considerable controversy. One particular issue is what we should call them. Every British micromoth – indeed, every species ever scientifically described – has a two-part Latin name. But fewer than one in ten of wee lepidopterans has a current English vernacular name that is sufficiently widely accepted for Phil Sterling and Mark Parsons to feel comfortable using it in their book – and most of those, depressingly, are for moths deemed in some way pestilent. This wasn't always the case: until the mid-nineteenth century, moths of all persuasions were granted common monikers alongside scientific. The English names fell out of favour because, Peter Marren suggests in *Emperors, Admirals and Chimney Sweepers*, they commanded less respect in the 'newly scientific pursuit'.

For many moth-ers, Latin exclusivity causes no problem. With effort and perseverance, anyone can learn a micromoth's name, whatever the language. But it is undeniably an additional entry barrier to an already challenging niche pursuit. Speaking only in foreign tongues risks making micromoths the elitist preserve of a modest number of scientific experts. 'Many [people] have trouble learning, remembering, spelling and pronouncing scientific names,' explains Jim Wheeler in

Micro Moth Vernacular Names, a 2017 book that sought to reinstate common names for smaller moths.

'Common names,' Matt Shardlow, boss of insect-conservation charity Buglife, observes in a Twitter discussion, 'definitely make it easier to teach people about other species and then conserve them.' This matters because we will only save the things we care about – and knowing them by name is a first step towards love. In an editorial for *Atropos* magazine, Mark Tunmore suggests that 'if we want people to care about the fate of a tiny micro-moth … it helps to be able to give it a name which has resonance with those less knowledgeable'. In *Our Place*, his treatise on the state of Britain's nature, Mark Cocker argues that environmentalists have long 'used language to ring-fence their profession' and that the exclusivity of scientific names hinders understanding for the layperson. Cocker's proposed solution – giving every species a common name – would be easy and effective: 'probably the biggest low-cost change' that environmentalists could make.

Intellectually, the battle seems nearly won. There is general acceptance that vernacular names for all micros would be sensible. The problem is what those names should be – and how they should come into being. Jim Wheeler's initiative is courageous and timely, but – like many a frontrunner challenging the status quo – he has taken flak for many names chosen. Nobody currently uses Wheeler's suggestion of 'Gold-base Tubic' for *Oecophora bractella*, for example. 'Common Tubic' seems an opportunity missed to enthuse people about the multicoloured wonder that is *Alabonia geoffrella*. Perhaps criticism is unsurprising. Peter Marren writes that

Wheeler drew extensively on monikers suggested by Ian Heslop in 1947 that either were scorned by specialists or generated hilarity. Either way, they were steadfastly ignored.

Mycologists resolved the absence of common names for fungi by arguing them out in committee. That is one possible route for micromoths. But in what may transpire to be an influential blogpost, academic Douglas Boyes argued in January 2019 that common names 'should be achieved, not assigned'. He favours an 'informal, open dialogue', starting by inventing and sharing nicknames for moths. 'If they resonate,' Boyes reckons, 'they might just stick.' Common names, he believes, should come into being organically, democratically. They should be 'an honour bestowed by the people'. I bet Geoff would agree.

'Is *Bisigna* on your "target list"? If it is, I've got one. If not, it should be.' The message comes from Nigel Jarman, a Kent moth-er and rocker. *Bisigna procerella* is indeed one of two micromoth goals for an early July trip to Britain's southeasternmost county. Putting something on a wish-list is easy; seeing it is a different matter. This is particularly true of *Bisigna*. First recorded in Britain during the infamous drought of 1976, *Bisigna* is still only known from relatively few sites in Kent and East Sussex. Trying to find it could be castles in the sky. This is why Nigel's message, received amid an M11 standstill, prompts whoops. Two hours later, we reach Nigel's home in Kingsdown, sipping a brew and cowering from a fly-over

Spitfire while admiring this tiny gem – the first he has caught in his garden, a hundred metres from the sea.

The *Bisigna* is drop-dead gorgeous. Being a fellow member of the family Oecophoridae, this half-centimetre-long wonder graces the field-guide page already bejewelled by Geoff and *bractella*. In shape, *Bisigna* recalls Geoff, albeit with less exuberant palps. Its predominantly tangerine form twinkles in the sunlight racing through Nigel's kitchen window. Three neon strip lights plunge vertically over the moth's wings, whose embers have burned charcoal smudges towards their tip. There's just one problem: it is titchy. Doing justice to its splendour goes well beyond reading glasses. It demands either a hand lens or taking a photo and admiring that rather than the creature itself. Either approach involves an interface that creates distance between observer and observed. It is not just scientific names that impede engagement with micromoths.

For the final diminutive quarry, we travel an hour south to the Dungeness peninsula. *Cynaeda dentalis* is a rare denizen of vegetated shingle and coastal chalk or limestone at scattered south-coast locations. Dungeness Bird Observatory warden David Walker is confident we will see it. We do so even more quickly than expected: as Will raises his foot to cross the Observatory threshold, he spots a *dentalis* resting on the brickwork by the door. Target nailed before even entering the building.

Cynaeda dentalis is one of those moths that cannot possibly be real. Rather than a paroxysm of colour like Geoff, *bractella* or *Bisigna*, the *dentalis* riot encompasses pattern. A complicated interplay of cream, tan and chocolate spikes and bristles creates a magic-eye optical illusion that denies the presence of invertebrate. This is not a moth but a grass seed. Or a miniature wheatsheaf. Or a tiny Hedgehog. It would take a canny bird to discern *dentalis* as food. This fingernail-sized creature quickly becomes one of the year's favourites. Eventually catching a hundred or so between Dungeness and Dorset, Will and I give it our own nickname: Hedgehog Moth. We should try it on Douglas Boyes.

The evening of our *Bisigna/dentalis* day sees Will and I prepare for a night mothing in Orlestone Forest, half an hour north-west of Dungeness. In this famous, oak-rich woodland, we will be joined by James Hunter, the generous, baseball-hat-wearing soul who coaxed me into moths, and Jac Turner-Moss, the Observatory assistant warden. Catching our own *Bisigna* is among the night's goals, so we ask a stony-faced David Walker whether he would want us to bring one back.

'No thanks,' he replies impassively. 'But I've never seen *Alabonia geoffrella*, so please could you catch that instead?' We giggle dismissively. Geoff's flight season ended a fortnight previously. What's more, we have never heard of this day-flying moth being trapped at night. David's wish is a non-starter.

Our Orlestone traps seduce several *Bisigna*, which delight us no end. Encountering our own moths propels us to a different emotional ballpark to seeing one temporarily imprisoned in someone else's pot. Over the coming weeks, many people experience their own *Bisigna* as this rapidly spreading species experiences its most fulsome year yet.

Will and Jac then trump the *Bisigna* with a truly leftfield moth. Grinning manically, Will recalls Keith

Kerr's adage from April. In July and at night, Geoff may be improbable but it is not impossible. The pair return from a trap round bearing a pot housing David Walker's long-craved moth. Jac will be the boss's favourite after this. Except when he peers inside the pot, Jac realises that Geoff is no more. The moth's final act, well after its season's sell-by date, was to flutter towards the light marking the stairway to insect heaven. Jac is crestfallen, visibly sagging with dismay.

'Don't worry, lads,' James Hunter drawls from a folding chair, where he is calmly examining moths extracted from other traps. 'I've got another one here.' Inside a pot, the night's second *Alabonia geoffrella* sits quietly. Lightning has struck twice, and this moth is alive and kicking. Come the morning, it may even coax David Walker into a smile.

By rights, that should be that. Four exquisite micromoths successfully tracked down, admired, eulogised and released. But Kent, James Hunter and tiny winged creatures haven't finished with us yet.

When flicking through the micromoths book to decide my year's goals, one emphatically patterned creature buckled my knees. Plonked unceremoniously at the foot of a page packed with lookalike brown jobs was a garishly yellow and red sweet that screamed icky E numbers. This was a moth like no other. Flicking excitedly to the text, I was distraught to learn that it was also no longer British. *Hypercallia citrinalis* had once occurred in Kent, and before that in Essex and Durham, but was now extinct. Its sole remaining outpost in the British Isles was The Burren in County Clare, Ireland. For a book restricted to the island of Britain, *citrinalis* was out of scope.

Fast-forward to late June. Will and I are slogging north towards Oban for several days' mothing when I spot an email from James Hunter. He wants to check that I 'know about *citrinalis*'. My heart races. Know what?

'Since news went quiet there have been counts of nineteen and six on different days at the site.' What news? What site? My mind is whirring.

'WTF?' I respond, in blind panic.

With my attention to social media suffering collateral damage from the bombardment of summer's incessant mothing, the astonishing rediscovery of this moth has passed me by. James explains that a botanist, Gareth Christian, photographed *citrinalis* on a steep grassy paddock concealed within Kent's North Downs, then alerted Kent moth-ers via Facebook before the details were hushed up to deter collectors, who might steal and kill the moths, potentially eradicating the newly discovered population. Chomping at the bit, James visited the site first thing the following morning and immediately chanced upon a *citrinalis*. In subsequent visits, he and a bunch of trusted friends found many more. Forty years after the last sighting, long after it had been logged as extinct, *citrinalis* was back!

Will and I, though elated for the moth and James, are otherwise gutted. We had been in Kent the previous week and could easily have paid homage. As it is, we must survive eight days on tenterhooks until we can go back. By the time we visit, nearly three weeks have elapsed since Christian's discovery. There is a real prospect that this moth's flight season is over, and that we will have missed the year's seminal moth event. We arrive early evening, perspiring up a steep hill past cranky yews and skanky youths performing BMX tricks. Throughout a sunny hour we contour lumpy chalk grassland infiltrated by scrub. We see plenty of Chalk Milkwort, the *citrinalis* foodplant. But of the moth itself, nothing.

In desperation, Will heads deeper into the sun-baked paddock. Wilting in the heat, I seek solace in a shady area where, unpromisingly, scrub has taken hold. Yet it is here that this contender for Britain's rarest moth wafts up from my toes and splays itself on a bramble sprig.

'Will! Will! Come here! Now!'

Hypercallia citrinalis is even more scintillating in real life than in the field guide. A mustard-yellow backdrop is etched, in violent strawberry-red, with a scowling voodoo mask. The figure's furious eyes pursue me as I reposition to take photos. The moth itself is unbothered by the fuss, flaunting itself in the open. Just as it presumably has been during each of the forty summers since its supposed extinction. *Hypercallia citrinalis* has been hiding in plain sight throughout.

How little we know about Britain's natural treasures. Particularly when they are small, undeservedly ignored micromoths.

8

Dry zone

Norfolk, Suffolk, Cambridgeshire and Kent
May–June

June opens its account in Breckland with me traipsing bemusedly along a weedy, six-metre-wide public bridleway through a vast, chest-high field of immature, lime-green wheat. This East Anglian monoculture does not immediately seem to promise an immense roll-call of moth species, yet Butterfly Conservation's Sharon Hearle is adamant about bringing our group of volunteers here. She should know. Nobody is doing more than this kind, measured environmentalist to save one of Britain's rarest moths, Grey Carpet.

The Brecks is a landscape unlike anywhere else in the country. Straddling westernmost Norfolk and Suffolk, its almost four hundred square miles of uniqueness derive from the interplay of climate, geology and changing human activity. Ensconced within a region receiving barely half the average rainfall for Britain, Breckland features a continental microclimate of warm summers, cold winters and aridity. Chalky soils are overlain with ancient patches of wind-blown sand, upon which a mosaic of lichen-rich, chalky and acidic grassland prospers. These grass-heaths are at their purest when sparsely vegetated and fractured by bare-ground microhabitats. Perversely, the more degraded and barren the land appears, the richer it is for wildlife.

And it is quite some bounty. The Norfolk Moth Survey will record 338 species of moth on a single night at a Breckland site this year. And a 2010 audit of Breckland biodiversity revealed just shy of 13,000 species breathing or growing in half of one per cent of the country's land area, many of which are distinctive rarities dicing with national extinction. The UK distribution of more than seventy species wholly or primarily comprises Breckland; seven occur nowhere else in the world. Save the Brecks and you protect a swathe of our rarest wildlife.

But saving Breckland is complicated. Across six millennia, our ancestors made surprisingly rich and unwittingly sustainable use of this spartan terrain. Low-intensity management involved rabbit-warrening, flint-knapping and sheep-droving. Soils were easily worked but poor. Following harvesting, early agriculturalists left fields fallow for years – so-called 'brakes' that inspired the region's name. This approach worked for wildlife. Rotation allowed natural vegetation to rejuvenate. Nibbling sheep kept dominant plants in check. Rabbits served as land engineers, creating bare ground that was critical for pioneer plants, the invertebrates that depend on them and nesting Stone Curlews – Breckland's 'wailing heath chicken'.

Messing with this long-established system broke the Brecks. In the late nineteenth century, sheep-graziers and their mobile mowers abandoned the region. Within scant decades, cultivation intensified and forestry plantations suffocated what was left – a deleterious transformation driven by public policy. In the 1960s, venerated moth-er E.B. Ford lamented in *Moths* that public authorities were exhibiting a habitual disregard for the countryside that, here in the Brecks, was manifest in their insistence on either afforesting the land or converting it to aerodromes and agricultural terrain. By the 1990s, three-quarters of the Brecks' grass-heaths had been destroyed, the remaining fragments disconnected, the landscape hostile to dispersing wildlife.

There has been no let-up in threats since. Buckled into the commuter belt for Cambridge and Norwich, incessant housing developments augment peri-urban sprawl and increase the human toll on wildlife. Pig farms, fertiliser run-off and vehicle emissions lay down ever more nitrogen, wrecking the lifestyle of poor-soil specialist plants, which become crowded out by swift-growing generalists such as nettles. Predation, persecution and disease have eviscerated populations of rabbits – ecosystem architects, the cheapest tool in conservationists' kit – by up to 96 per cent. Milder winters with increased rainfall – partly due to 20,000 new hectares of plantation forests increasing humidity and buffering temperature extremes – emasculate microclimate continentality.

Such changes have levied a catastrophic cost on biodiversity. Fifteen Breckland species have become nationally extinct. Nearly half the disappeared are moths. Their number includes two former Breckland exclusives, Spotted Sulphur and Viper's Bugloss. Judging from photos, the first-named is magnificent, with tented primrose-yellow wings striped black like a tiger and spotted yellow like a leopard. The last British example was seen in 1960. Viper's Bugloss – a relative of Lychnis, a common summer moth – is ribbed and pockmarked with tan, chestnut and black. The final British record was of a caterpillar in 1968 – a lonesome larva that, even if it had survived into adulthood, could not have saved its species. Its common name suggests an ecological association with the eponymous, stately indigo-flowered plant that poses proudly and commonly across Breckland. The assumed link was erroneous; the caterpillars fed on Spanish Catchfly, a very rare plant. Sadly, by the time Viper's Bugloss was afforded legal protection through the 1981 Wildlife and Countryside Act, it had already been extinct for an unlucky thirteen years. That clunk you just heard was the sound of the stable door shutting.

Although E.B. Ford did not explicitly presage this duo's extinction when he specified them as two of Breckland's three most special moths, he would have been acutely aware of their dwindling populations. Openly pessimistic about the region's fate, Ford judged that whatever tracts of land might be saveable would almost certainly be insufficient to safeguard the Brecks' cherished characteristics. It seems fitting, then, that Sharon Hearle has assembled our group today to search for the third Breckland moth that Ford mentions.

Beneath heavily ruffled, gunmetal skies, a dozen of us plough along the overgrown bridleway. Before us, grasses and herbs tuft up in greeting, among them frayed threads of yellow-flowered Flixweed. The seedpods of this arable 'weed' are manna for Grey Carpet larvae. The plant's abundance delights and surprises Sharon: 'We're lucky that there's so much Flixweed here, given that it's sandwiched between wheat rather than sugar beet.' Sharon wears her expertise lightly. She explains that crop-specific pesticides are the problem, rather than the crop itself. The chemicals drenching wheat eradicate Flixweed, without which Grey Carpet stands no chance.

Both plant and moth would have thrived in yesteryear's 'brakes' and rabbit-burrowed terrain. Today they are condemned to live on the edge, in field margins and fallow plots, along roadside verges and on disturbed ground. They are squatters, travellers perpetually forced to move on.

Grey Carpet's intimate relationship with bare ground becomes evident a hundred metres down the bridleway. A fragile-looking, whitish moth flutters from a spray of Flixweed then down to the ground, where it evaporates. Unusually among moths, Grey Carpet's camouflage does not derive from disruptive patterning. But the dusting of black dots over dirty-white wing-triangles serves just as well on this fine, stony soil.

The moment I manage to discern the moth's outline, it flickers upwards to land on a wheatsheaf. The ironic juxtaposition of rarity and threat escapes none of us. According to the seventy organisations that produced the 2019 *State of Nature* report, agriculture is the overriding cause of UK wildlife loss. This is unsurprising, given that the industry manages nearly three-quarters of UK land, but universality is no excuse for damaging practices such as environmentally wanton use of chemicals that disinfect nature. Conservationists have long observed that recent declines in Western and Central Europe's butterflies and moths correlate strongly with agriculture's increasing intensification. In 2018 researchers led by Susanne Kurze evinced that the more plants were enriched with nitrogen (from fertilisers), the higher the death rate among moths.

As the twenty-first century started, only a handful of locations still held decent populations of Grey Carpet. A re-run of the fate of Viper's Bugloss and Spotted Sulphur was a realistic possibility. It was known that Flixweed, and thus Grey Carpet, required regular disturbance of the soil surface to take root. The same was true for Breckland's other star moths, including Marbled Clover, Forester, Tawny Wave, Basil-thyme Case-bearer and Lunar Yellow Underwing, whose caterpillars I had helped Sharon survey in January. Soil disturbance, Sharon mused, could be arranged – through mechanical rotavation, turf-stripping and other techniques. Together with Butterfly Conservation colleague Sam Ellis, Sharon literally and metaphorically broke new ground by creating fifty-nine bare-soil plots from scratch. It worked: target moths appeared at half the plots, Flixweed on a third and Grey Carpet at one in eight.

Sharon has been spreading the knowledge that Grey Carpet does well on cultivated arable margins where the ground is disturbed each year, and has persuaded highways contractors to rotavate broad verges flanking the A11 dual carriageway, which guillotines Breckland. As we drive the road to join Sharon, we spot Flixweed, Common Poppy and other arable plants. Sharon wants the contractors to integrate rotavation into their verge-management regime. It will be cheaper than regular mowing and will refresh bare ground that Flixweed can colonise. She is confident that Grey Carpet will follow. 'For a weak-looking moth, it's good at dispersing,' she says.

Altogether, we root out seven Grey Carpet on the wheat-field Flixweed. This bodes well. Tragically, when Sharon returns weeks later, the overgrown bridleway has been shorn – its Flixweed, and perhaps its Grey

Carpet too, paying the ultimate penalty for being in the wrong place. Our abhorrence of 'messiness' denies wildlife a home.

This spring and summer, I shun work to luxuriate in several half-days shambling around Breckland. Its landscape is deceptive, barren flatness concealing considerable bounty. I attend moth-trap openings at the Norfolk Wildlife Trust reserve of Weeting Heath and at Grime's Graves, a Neolithic flint mine managed by English Heritage. At the latter, an event run by Sharon Hearle, tourists seem as intrigued by the moth catch on display as they do the 5,000-year-old remains. Their exclamations convey agreeable surprise, even hinting at enchantment.

'So none of these moths eat my clothes? Just plants?'

'I never knew moths were so colourful. They're prettier than butterflies.'

'If they're not dead, why aren't they flying away?'

'I'll never be scared of moths again.'

'That is *gorgeous*!'

After January's nocturnal tummying, I also visit Cranwich Camp during daylight. Only now can I properly appreciate the bare-ground quadrangle that Sharon created. It has become littered with Spanish Catchfly. Given how well this spindly botanical rarity has recovered, some moth-ers are mooting the reintroduction of Viper's Bugloss. Today, the only Viper's Bugloss I see are the plants, which clump proudly across the vista. At one, I spot a Marbled Clover nectaring – buzzing like a

Hummingbird Hawk-moth as it zips between flowers. In flight, it whirrs caramel and humbug. When the moth alights, my eye is drawn to its strongly banded hindwings. Even more striking are the Foresters. Garbed in Lincoln Green cloaks, they stand sentinel on prominent fronds of grass, their alert demeanour emphasised by attenuated, vigorously feathered antennae.

One stifling afternoon near Lynford, I mount a gentle incline from whose stony base thrusts a floral riot. To the percussion of humming bees, I chance upon a Tawny Wave. This East Anglian speciality is the size and shape of a Grizzled Skipper butterfly but regales with its strong rosaceous hue. Another day, at Santon Downham, I trawl forest rides and undulating heathland. In the seventeenth century, Breckland sands were so mobile they formed inland dunes that almost engulfed the local village. I pay my respects to Perennial Knawel, a botanical speciality of broken-ground Breckland that has been reintroduced here.

This tiny, prostrate plant and four moths – including Grey Carpet – are among sixteen key species designed to benefit directly from Shifting Sands, an appropriately named conservation project in the Brecks that forms part of the Back from the Brink portfolio. This audacious initiative assembles conservation bodies across England into partnerships to rescue a score of near-extinct species and help ten times more. Its unparalleled ambition and early successes will later justify Back from the Brink scooping top prize for heritage projects in the National Lottery's twenty-fifth birthday awards.

These endeavours provide succour during my solitary wanderings. As I stroll, I am mindful of local moth-er

Graham Geen's calculations, published in the magazine *British Wildlife*, of sweeping modifications to Breckland's moth assemblage – driven by searing changes to its landscape and climate. At first blush, the subsequent and overall figures of comings and goings look surprisingly welcome. Since 1980, 362 species have been recorded for the first time in the Brecks compared to 135 that have vanished. But the additions are largely widespread generalists that could make a home anywhere – and the losses principally specialists for whom the bell may be tolling. Twice as many of the very rarest species – those listed in the Red Data Book – have vanished as have arrived. Can Shifting Sands rescue the Brecks before they rupture irreparably?

When we first met in January, Sharon Hearle namechecked another sad addition to Geen's litany of Breckland absentees. Four-spotted last bred here in the mid-1990s, a sorry fact ascribing greater resonance to its scientific name – *luctuosa*, meaning mournful. At the turn of the twentieth century, moth authority James Tutt commented matter-of-factly in his *Practical Hints for Field Lepidopterists* that 'most country lovers' were familiar with it. A denizen of weedy field margins, Four-spotted was seemingly a standard part of rural life south of a line connecting the Humber and Severn estuaries, yet it has since been ousted from much of its British range. Within a century, Roy Leverton wrote in *Enjoying Moths* that Four-spotted had suffered 'one of the most dramatic changes in status of a moth that has not yet become extinct'. As agriculture intensified, Four-spotted numbers plummeted.

Nature enthusiasts rightly think of crop-shielded landscapes as usually being vast wildlife deserts – drenched in pesticides and shorn of natural vegetation. But it is not always so. Even tiny remnants of natural plant life potentially harbour unusual creatures. Sharon encourages me to look for Four-spotted amid Royston's pseudo-prairies, which monopolise the intersection of Cambridgeshire, Essex and Hertfordshire – just 35 miles south-west of Elveden. Here during 2018, Sharon bumped into the moth in several new places, and she is keen to understand whether that year's heatwave has strengthened its populations.

Ben Lewis, Wingman Will and I accept Sharon's challenge, surveying a series of sites around Royston. Four-spotted flies by night and, if the conditions are calm and sunny, by day too. As we pull up on a concrete pad just off the A505's busy cross-country thoroughfare, blustery squalls rip across the expansive wheat-fields. Our hopes plummet. Even if the moth does occur in this most unprepossessing of locations, it won't be airborne today. The following two hours see only the occasional let-up in wind; glimpses of sun prove yet more sporadic.

But we are not thinking like moths. To do so, we descend to ground level, scrambling down into a two-metre-wide ditch separating field from road. Kneeling, my eyes lie lower than the asphalt, enables me to enter a microcosmos of Weld and poppies and Ox-eye Daisies and much more greenery besides. With clumps of field Bindweed – the moth's larval foodplant – knotting steep, bare slopes fronting the early afternoon sun, this is plausible Four-spotted terrain. And so it proves. We count thirty-eight here and a further seventeen across two other nearby field margins. In inclement weather, we cannot believe our fortune. I text Sharon; she is over the moon.

Four-spotted is a skittish creature, cautious of our gargantuan forms, usually preferring to fly a few metres rather than hide in plain sight. Its flickering flight tricks the eye until it vanishes. When grounded, its dress blends with the bitty terrain of admixed flint and sundered, chalky soil. Funereal wings are indented with four prominent rosy-white blots that disrupt the moth's overall form. Even after watching one alight on the road's pebbly fringe, it takes me a full minute to clasp eyes on it. Another Four-spotted whisks up into an adjacent, luxuriant Hawthorn hedge. It proves to be a gravid female, her belly plumped up with eggs.

The ditch of joy, Sharon says, was dug by the farmer two years previously to prevent the ingress of hare-coursers. A decision taken to help a mammal has brought unexpected additional benefit. 'The bare ground,' she explains, 'enables Field Bindweed and other arable plants to establish, providing better habitat for Four-spotted than grassy field margins and road verges.' Leaving roadside ditches seems to be a welcome local practice. Sharon is working with farmers to ensure their continuation. Ideally, in order to rejuvenate the arable annuals, she'd like cultivators to scrape some bare ground each year.

With friends like Sharon, the prospects for Four-spotted may not be as bad as Roy Leverton feared two decades previously. Indeed, this transpires to be an exceptional year for the species. Not only do known sites harbour the highest numbers for years but this delightful moth also appears at several locations with few or no previous records. Would it be rude to conceive of Four-spotted's return to the Brecks?

Although it primarily favours the margins of life, Four-spotted is – or was – also a moth of chalk downland. This soft, porous rock exerts a strong influence on the aridity of both Breckland and Royston's agricultural plains, but reaches its pinnacle in southern England's roller-coaster downs. To experience this dry zone's day-flying moths, Wingman Will and I scurry to Kent, a county whose moths are so special that the National Lottery Fund is supporting their conservation.

Lucia Chmurova is a twenty-something entomologist who, when we meet, is managing the early stages of 'Kent's Magnificent Moths', a Butterfly Conservation project. With exuberant earrings and punk-dyed hair, Lucia punctures preconceptions of what an insect-focused biologist should look like. One of a cadre of young, talented female naturalists currently energising Britain's conservation scene, Lucia cut her teeth on decades-old beetle collections hidden in the Natural History Museum's attic, then fell in love with tropical rainforests through researching Bornean bugs. Now she is saving British moths.

'The Kent's Magnificent Moths project has two main strands: conservation and engagement,' she explains. Kent is home to the UK's greatest concentration of threatened moths – but few residents are aware of their existence. Lucia wants the project to change all that. 'We are focusing on eight very rare moths – some so scarce that we don't understand enough about their population trends to coherently explain the urgency of their conservation to potential funders. And we want to excite local people about moths more widely.' To help protect and enhance the conservation status of moths such as Bright Wave and Fiery Clearwing, she continues, 'we are working with as many partners as possible – from Kent Wildlife Trust to golf courses and farmers. Some people help out of the goodness of their heart, but others need to be persuaded of the benefits to them.' There are even school and community group 'champions' for particular moths. The project seeks to use these underdogs of the animal kingdom to surprise, inspire and raise awareness of the diversity of nature, and opportunities to engage with it, she explains with mounting excitement. 'We've shown people moths in a city-centre park, a suburban garden and alongside a kids' pond-dipping session. We've targeted hard-to-reach groups, such as teenagers and people with mental-health issues. Their response has been amazing. It is so exciting – and surprisingly easy – to change perceptions about moths through first-hand experience.'

One of Lucia's volunteer consultants is cheery local moth-er Ian Roberts. He invites Will and me to join him at his stomping ground of Abbot's Cliff early one fine June morning. Midway between the seaside bustles of Dover and Folkestone, we meet near a concrete 'sound mirror', an inter-war listening device intended to warn of incoming enemy aircraft. Standing proud at the head of the high white cliffs that fashion Dover's fame, the four-metre-tall concave structure faces off at the Continent.

Ian's interest in moths started on teenage visits to nearby Sandwich Bay bird observatory, where the winged insects have been studied for decades. He returned to the hobby sixteen years ago, and now 'traps pretty much every night'. Although primarily driven by learning more about scarce moths – a prerequisite for protecting them – he is thrilled by showing their common brethren to the general public.

'If people even get beyond assuming that every moth eats their clothes, they think they are all small and brown. So the range of colours really astonishes them. So too the variation in size: from a tiny five-millimetre-long micro to a Privet Hawk-moth twenty times bigger. Above all, people are amazed by the diversity. You try showing people thirty-five types of butterfly on a single day! And thirty-five species is not even a good night's mothing.'

We have joined Ian to see a creature that narrowly missed out on inclusion in the Kent's Magnificent Moths project. Dew Moth is an oddball. It is a footman but not named as such. Unlike other footmen, males readily fly by day. In Britain, it is rare and declining with only occasional far-flung dots remaining on the map. I have previously watched Dew Moths above 3,000 metres in Spain, yet in Britain the species inhabits rocky coasts and chalky grassland. In some places, its larvae are almost seafarers, feeding on wave-splashed lichen on rocks only marginally above the high-water mark. In Kent, Dew Moth thrives in a warm, dry climate; in Scotland, it relishes mild, damp conditions.

We see them easily, as the moths flush at several metres' range then career around before plonking down in hummocky grass. After a minute's sulk, most crawl up a grass stem to resume their lookout, suspended like a drop of molten caramel. Dew Moth proves a lovely thing, a fey Imperial stormtrooper wearing a leopard-skin cloak that covers a yellow-tipped black abdomen. We count fifteen, all within feet of the friable cliff edge, below which the caterpillars nibble lichen.

'This population is doing well,' Ian says. 'That said, I've counted no more than thirty this year compared to 150 in the past. Amazingly, the first emerged in March, which is earlier than ever. They used to fly only in June and July. Things are changing.' With thoughts of the impact of the climate crisis on wildlife seasonality, I peer cautiously over the cliff. I wonder how many seconds it would take to tumble to the inky rocks below. I inch backwards. We have other moths to see.

Driving west beyond Folkestone, we glance upwards at a searing chalk slope that boasts a purview over the Channel Tunnel entrance. Later in the summer, we will slide around Cheriton Hill's long, blowy grass and shamefully artificial white horse, looking for one of Kent's magnificent eight: Straw Belle – a dun-coloured equilateral triangle – achieved modest fame in adjacent Surrey during the 2012 Olympics. The zigzags of Box Hill were a challenging section of the Olympian cyclists' road race, but also adjacent to an important population of Straw Belle. Conservationists feared for the moth's sanctity, but environmentally sensitive organisation ensured that 30,000 spectators watched furious pedalling without harming moth or habitat.

Turning our eyes southwards, we glimpse another downland ridge. This one harbours Kent's last remaining population of Four-spotted, a moth just about clinging on here. It would be fun to call in, but the morning is pressing us towards an even rarer moth. Within an hour we are huffing up the base of the Devil's Kneading Trough, a steep-sided coombe that incises Wye Downs.

Broiling summer days such as today excel for wafting around chalk downland. Far above us, at the top of the escarpment, two stick figures whirl a crimson kite, while lovers share a moment astride a bench. On the steep, sheep-runnelled slopes beneath our feet, our attention halts upon the candy-pink of Chalk Fragrant-orchids and stately, ivory-flowered Greater Butterfly-orchids. Then our tread and gaze return to the waist-high tor-grass that brushes the valley bottom. If we are going to see a Black-veined Moth – the poster child of Kent's Magnificent Moths – it will surely be here.

And so it is. Initially, I dismiss it as a Small White butterfly, which is of similar size, shape and pallor. Only the delicacy of its billowing flight makes me look twice. We observe the moth discreetly through binoculars to avoid disturbance, because Black-veined Moth is among eight über-rare (and, in one case, extinct) moths swaddled in legal protection. Fortunately, distance does not impede us easily admiring the bold black cell-edging that drives the moth's common name. It is a new moth for Will, and a broad smile fissures his beard.

Known historically from thirteen ten-kilometre squares sprinkled across southern England, Black-veined Moth now occupies just two such squares, both on Kent's North Downs. It could actually have been much worse. In the 1980s, the moth flew at only two sites, teetering on the very cusp of absence. Since that nadir, Natural England, Butterfly Conservation and others have strived to revert 200 hectares of arable land across twenty farms into long-turfed, herb-rich downland. It is, Kent conservationist and moth-er Sean Clancy says, 'the ultimate problem species for which to manage' as its needs are precise and could be deemed contradictory when trying to artificially maintain or create suitable habitat. 'Black-veined Moth requires mid-successional grassland with a healthy, species-rich sward alongside established, dense, grassy tussocks to provide larval cover in winter,' he explains. 'This is a difficult combination to achieve.' Too many grazing animals – or too few – and you lose or miss the transitional stage that the moth needs. 'We divide sites into compartments with different grazing regimes, but it can be hard to get farmers to understand what's needed.'

Even though the moth now spreads across eight locations, its population slumped 60 per cent between 2002 and 2016. This species remains in critical danger of national extinction, needing all the help it can get. Yet the moth's conservation is not without controversy. Botanists have expressed concern that the grazing regime designed to help Black-veined Moth – which encourages growth of long, dense tor-grass – has choked rare plants such as Burnt Orchid and Late Spider-orchid. With modest tinkering, it should surely be feasible to resolve such conflicts.

No such dispute envelops *Anania funebris*, the only micromoth to be granted membership of Kent's octet. This is another odd moth – one of a select band that inhabits North America as well as Europe. For Butterfly

Conservation's Steve Wheatley, *funebris* – sometimes known as White-spotted Sable – is the flagship for nine moths feeding exclusively (or nearly so) on Goldenrod. This perennial herb with fluffy yellow flowers grows in woodland glades on chalk and limestone, but is declining in lowland Britain, probably because it is getting shaded out. One Goldenrod-feeding exclusive, Cudweed, is already extinct in Britain. Steve worries that others, including *funebris*, may follow.

In an empty segment of the map near Canterbury, Will and I make for an orchid-strewn enclave hidden deep in a shady, aqueous wood. A Nightingale kicks off – one of only two examples of this dramatically declining songster that I hear all year. For half an hour, there is no suggestion of anything but uninterrupted canopy, but suddenly we open out on to a secret garden of chalk-flecked turf, scattered bushes, space and sky.

I spot a *funebris* shortly after arrival, but Will is too entranced by the display of Lady Orchids to see it. We wander and wait for two tense hours before we glimpse another. This is a winsome moth, neater than the Four-spotted that it superficially resembles. Each sooty wing is circled twice with whiteness, and golden pigtails whoosh backwards from the moth's head. It flies as would a conjurer, smoking upwards in a flickering spiral before dissipating into thin air. We wander and wait some more, but the clandestine *funebris* has divulged enough of itself for one day.

As we depart, I realise that five of Kent's eight 'magnificent moths' happily fly by day, two exclusively so. In her essay 'The Death of the Moth', Virginia Woolf argued that 'moths that fly by day are not properly to be called moths' for 'they do not excite that pleasant sense of dark autumn nights and ivy-blossom'. Woolf is taxonomically incorrect, of course; *Anania funebris* is as much moth as an Angle Shades. But she is taxonomically astute too, her inference being that we should regard day-flying moths as we do butterflies – and the latter are nestled, evolutionarily, within moths.

The surprising prevalence of day-flying moths on chalk downland proves entirely explicable. Once the sun has absconded, the escarpment's convex slopes – only sparsely insulated with scrub – briskly radiate heat. Even after sweltering days, nights can cool so rapidly as to render moths inactive. Totting up, I realise that only 10 per cent of British moths are habitually day-flying, but this still means we have easily four times as many types of day-flying moth as we do butterflies. Nevertheless, public perception unjustly tars all moths as having 'something of the night' about them, banishing them to the realm of the unwashed, the pestilent and fearsome. Perhaps Kent's Magnificent Moths will change the record.

9
Wetsuit

Norfolk, Lincolnshire and Cambridgeshire
May–July

A thousand times more people visit the Norfolk Broads each year than live there. Most journey in search of wilderness – a deep irony given that this East Anglian waterland is a surreptitious human construct derived from natural-resource exploitation. But the wildlife doesn't mind. Harnsers (Grey Herons, in Norfolk speak) lope along dykes, butterbumps (Bitterns) stealth-hunt channels where polywiggles (tadpoles) clot. But most of all, there are moths, unseen denizens of the Broadland night, making more robust a living from fen than the county's remaining thatchers do from reed. It's time to make the most of living in Norwich, by seeking out some particularly special wet moths.

It was a redoubtable botanist, Joyce Lambert, who determined that the Norfolk Broads were, in large part, made by humans. In the decade after the Second World War, Lambert was intrigued by the sharp demarcations she observed between clades of vegetation. Digging into the underground peat revealed straight lines and flat bottoms – sure signs of human implement. Further research determined that Nordic settlers had started extracting peat for fuel by the tenth century. After four centuries of industrial exertion, rising sea levels swamped the cavernous pits left behind. In today's waterscape mosaic, blue-green reedbeds, scruffy fens, grazing marshes and damp woodland carr pixelate around stilled lakes bonded by 150 miles of waterways.

Freshwater wetlands reveal almost nothing of their moths by day. Interrogating the habitat is a task for bright lights on overcast, cloud-comforted nights that mitigate the inherent chill of open water. In his book *Moths*, E.B. Ford considered the moths of damp places 'an immense assemblage'. Broadland is renowned for possessing a suite of dampness-liking species that occur only rarely elsewhere, if at all – their distribution perhaps inhibited by the colder climate of other wet regions. Evolutionarily, wetland moths are also an intriguing set. Ford observed that many species from various branches of the taxonomic tree are both exclusive to these damp habitats and exhibit perfect colour adaptations for living among reedbeds. Not only are such moths the hue of an out-of-season reed stem, but their wings have raised veins that match the plant's texture. As Ford suggested, moths from different evolutionary branches have discovered the same trick, a concept known as convergent evolution. Reed Dagger, Fen Wainscot and Common Wainscot sit in separate sub-families, yet all exhibit the same fundamental guise that presages a fundamentally similar survival strategy.

Appropriately, it is a wainscot that forms the first Norfolk target. Very early on May's final Sunday, I accompany a band of undergraduates as they check moth traps set in secluded parts of the University of East Anglia's riverside campus, west of Norwich. I am welcomed by Josie Hewitt and Max Hellicar. The duo are intent on winning an inter-university competition run by Butterfly Conservation and a youth nature network, A Focus on Nature. The fourth University Moth Challenge seeks to encourage more students to formally record wildlife, particularly moths.

Josie and Max want to learn which moths frequent the campus, particularly its wetland fringes. But they seem particularly keen to win the category on participation – the highest number of people taking part in a single surveying event. I now understand why they invited me; I help make up the numbers. Impressively, Josie and Max have coaxed out six ecology students from their weekend hangovers, including two who have never previously seen a moth. Their number include an Ecuadorian, Marna, who admits that she snuck out without telling her flatmates what she was doing. 'I couldn't tell them I was going mothing,' she grins. 'They wouldn't understand.'

Josie recalls that she got into mothing because of its similarity to bird-ringing. In this branch of birdwatching, birds are temporarily caught and their legs marked with metal rings that, upon recapture, shed light on movements, longevity and ecology. 'But I *stayed* with moths,' Josie says, 'because mothing is fun and moths are cool.' She admits that starting out was overwhelming but that it 'gets easier once you can recognise a few species'. Above all, 'there's always more to learn and discover. I have seen 750 species in Britain – and there's twice that still to see.'

We head off the boardwalk, moistening our boots while examining a trap in a boggy fen that neighbours reed, birch and alder. Working by rote, Josie calls the name of each individual moth on a score of egg-trays tessellated within the black plastic tub, while Max diligently transcribes them. By the morning's end, the pair will have assiduously identified 700 moths of a hundred species. It is impressive work.

Pots closeting attractive or interesting creatures are passed a round for perusal. Poplar Hawk-moths provoke gasps of astonishment from the newbies, while everyone admires Scorched Wing, a bronzy moth burnished with griddle marks and glazed lilac. Josie eases a Pale Prominent onto her finger. It looks for all the world like a bark chip, the embodiment of crypsis – so much so that one newbie moth-er takes to Twitter to seek advice on whether it actually is a moth. 'Just look at it!' Josie cries. 'It's so weird, such random evolution – and just so cool. Moths are so much better than butterflies.'

But it is a Broadland speciality that Josie is really after. Hailing from Hampshire, where the species is absent, she is overjoyed to catch a handful of Flame Wainscot. Wainscots are so named because their pale, grainy wings reminded eighteenth-century Aurelians of the wood-panelling (wainscoting) in fine houses. Across the year I will see sixteen types of wetland-dwelling wainscot, though sadly not the rare White-mantled Wainscot, which we fail to find on our friend Alison Allen's marshland farm. Today's particular nationally scarce species is an East Anglian speciality, the Norfolk Broads its stronghold. Essentially straw-coloured, Flame Wainscot is delicately tinted pink ('flame' is over-egging it), while pale and dark blazes vie for attention towards the outer wing. Its shape is unlike any other moth, being slender and elongated with an arched wing tapering to a fine point.

By 10 am everyone is happy, buoyed by time outdoors, sparked by new sights and buzzing with insights. Even Marna – gazing adoringly at a hulking, furry and delicately inscribed Puss Moth that is luxuriating along her finger – is glad she came. 'It's fun!' she says. 'But I still can't tell my flatmates.'

Well to the west of the Norfolk Broads lie the remnants – a solitary 1 per cent of the original expanse – of what once were truly wild waters, the Fens. Britain's answer to the great Danube Delta of eastern Europe, this formerly vast wetland extended over nearly eight hundred square miles. Arcing around The Wash estuary, into which Fenland drained, it saturated west Norfolk, Cambridgeshire and south Lincolnshire. The mires and meres were inhabited by hardy, amphibious folk – pejoratively known by outsiders as 'breedlings' or 'splodgers' – who survived through harvesting sedge and reed, fish and fowl.

The Romans started it, of course. The most imperious of colonists were the first to banish water through dykes and embankments. But the most uncompromising campaign to rid liquid from land was led by Cornelius Vermuyden in the mid-seventeenth century, under commission from a local laird. Two centuries later, conversion was complete: dry land had been salvaged, its fertility reinventing the region as England's breadbasket. Today, writes Mark Cocker in *Our Place*, travelling by road through fenland roads often gives 'little sense of water, let alone of wetland'. Place names now provide the only remaining indication of water. Modern naturalists can only dream of the wildlife riches once coupled with Fenland. Breeding Common Cranes and Spoonbills, ample Ruff and Black-tailed Godwits, certainly. But what of the moths?

We never shall know for sure how bounteous were Fenland lepidoptera. By the time the Victorians arrived with their moth-trapping paraphernalia of sugar and paraffin lamps, there was barely enough time to catalogue species before they disappeared. Draining the Fens was the single most dramatic cause of British moth extinctions. All were wetland specialists. By 1879, the country had lost Many-lined and Reed Tussock. Species that frequented damp margins rather than outright wetlands enjoyed a few decades' grace. But by 1907, Gypsy Moth had vanished; by 1915, Orache was extinct. Marsh Dagger clung on until 1939, before joining the ranks of the disappeared.

To honour their passing, I pledge to track down some Fenland survivors – even if doing so means travelling outside that single remaining percentile of habitat. The first quarry is a capricious rarity, Marsh Moth – a great prize for Victorian collectors. Drainage extinguished Marsh Moth from Fenland during the 1960s. Before the species could be declared extinct in Britain, however, it was unearthed in Lincolnshire. Here it clings on today, albeit at just two coastal reserves, and is considered nationally endangered. As its name suggests, Marsh Moth's East Anglian population was associated with wet fens, where its caterpillars fed on Meadowsweet. Oddly, Lincolnshire's colonies approach life differently, frequenting the drier terrain of herb-rich dune slacks rather than adjacent marshes. Here it is really

Marsh-edge Moth, at best. Even more strangely, Lincolnshire caterpillars munch Ribwort Plantain, a plant so widespread and common that it sprouts from our garden lawn. Yet I don't have Marsh Moths at home. Meanwhile, there are rumours that females don't fly. This species is an enigma, for sure.

In a year-long quest to see rare and remarkable moths, failure constantly whispers its imminence. Bad weather, early flight season, late emergence, inept choice of trapping sites, elusive species and absence of permission unite to render the task uncertain at best. On a breezy late-May evening Will and I arrive at Saltfleetby-Theddlethorpe Dunes National Nature Reserve – a former RAF bombing range, gifted to the Nature Conservancy in 1969 – on Lincolnshire's soft coast. The welcoming party comprises skies heavy with rain and rushing cloud, doing little to boost our confidence. Common Blue butterflies have already hunkered down to roost, their goal simply to survive the storm. So it is with surprise and relief that we hear Lincolnshire Wildlife Trust reserves manager Matt Blissett make a confident proclamation. 'Don't worry,' he says, all easy-going smiles and approachability. 'We will see Marsh Moth tonight. It doesn't mind the cold.'

Joined by Northamptonshire moth-er Malcolm Hillier, we set up three traps in a dry football-pitch-sized area with ankle-height, herb-rich sward and two in a swampier area, where Matt caught seventeen Marsh Moths the previous week. Behind us, a male Cuckoo bangs on, demanding service. To the east lie hummocky sand dunes and resolute saltmarsh; beyond them, a distant dream on this accreting coastline, is the North Sea.

As well as managing this and other reserves, Matt has studied Marsh Moth for the past decade, counting adults in spring using lights and caterpillars in autumn by sifting through piles of leaf litter. His counts defy correlation: 2017 featured the lowest maximum catch of adults (seven) since 2011, but also the highest total of caterpillars (forty). The moth's ecology is similarly mystifying. Not only is it absent from suitable habitat, but it used to occur further north in the reserve until drenched by flood in 2000/01. 'When you think that it disappeared from Fenland because of drainage, it is strange that here we've experienced the opposite,' he says. 'The moth went downhill when conditions got wetter.'

In an attempt to restore balance, conservationists tried to move larvae to Saltfleetby-Theddlethorpe and establish a new colony there. The attempt failed. 'We just didn't have enough caterpillars – twenty rather than 2,000 – to make translocation work,' Matt sighs. Current hopes revolve around the recent discovery of a healthy population of Marsh Moth at a nearby reserve. Matt describes this as a 'eureka moment', long hoped-for, 'because we were starting to fear we had just one population to work with.' He explains that the Holy Grail 'would be to find a female. Long ago, a moth-er found one below his groundsheet, but that's the only incontrovertible British record I am aware of. No wild female has ever been photographed. Even the Victorians never collected one.'

This is quite the revelation: a moth so rare that, in Britain, only one gender is effectively known in life. I call over to the Wingman. 'Your task tonight is to find a female. They basically don't exist, but if anyone can make history, you can.' Where there's a Will, there's a way and all that. Sagely, Will ignores me. If the great and the good haven't found a female, he won't either. Shivering under our fleeces and waterproof trousers, we will be lucky enough to see a male.

Matt jerks the cord on his generator, cranking a hum into the air that drowns out the mechanical dusk serenade of a Grasshopper Warbler. Were the conditions still, we might hear the churring of Natterjack Toads. But tonight the wind roars.

In such conditions I am surprised that so many moths are flying. We quickly notch up perhaps fifty Small Elephant Hawk-moths – nearly tenfold more than I have seen across six years of mothing. Several female Fox Moths – chunky, chestnut dames – scatter eggs wantonly on trap wires and groundsheets as well as straggly vegetation. An enchanting Peach Blossom – velvet with orange and rose flowers, like Laura Ashley wallpaper – is marginally outclassed by the nationally scarce Sand Dart, a new moth for Matt and Will. As our hopes feint to rise, Matt cautions that Marsh Moth flies late. We should not expect one until the final half-hour before midnight. But that period comes and goes. The wind gusts, spirits sag. Then, forty-five minutes into the new calendar day, Matt calls out: 'Marsh Moth! Marsh Moth!' He points to a male sat just above the ground in the lee of his trap. We rush to greet it.

Granted, Marsh Moth is never destined to be a pin-up, its hues a greyer beige than a wainscot. Jagged lines across its wings track the share price of a particularly volatile stock. Subtle and understated, it is a moth-er's moth. Matt recalls showing one to a colleague the previous week. She had wanted to know what all the fuss was about. 'The look on her face,' Matt says, 'suggested she was thinking: "Is that it?"'

Eye candy it is not, but Will and I nevertheless exchange beaming smiles. Success against the odds.

Then the world goes mad. Matt spots a second male, not one metre away. Will spots a third male, even closer. Three males in a tiny area, all on the ground, all running around. This makes little sense. Then Will realises that the third male is circling the wiry stem of a Ribwort Plantain. Round and round it scuttles, as if following a scent. Running his eyes upwards, following the plant's modest height, Will's gaze alights on the brown, oval flowerhead, and he starts. 'Is that…? Is that…?' he stammers. 'Is that… a female?'

Poised atop the flowerhead is another brown moth – smaller, duskier, fresher than the male Marsh Moths. Will quickly surmises that this is a she-moth 'calling' – issuing a sex scent – that has lured in the ground-running males below. Bony fingers trembling, Will inches a pot over the moth so we can take a closer look. None of us can believe our eyes. Matt's professed Holy Grail is about two-thirds the size of a male, with markedly rounded wings that look capable only of rudimentary flight, if that. Her wing markings match the males', but she is darker, more compact. Matt's face is etched with ecstasy, perhaps with the barest tint of agony: this should have been his find, not Will's, but he is magnanimous: 'Find of a lifetime, Will. Well done.' Will is granted the first photograph of a wild, live female Marsh Moth in Britain. The rest of us waste no time following suit.

Months later Malcolm tracks down the story of the only other documented female. On 12 June 1948, at

Woodwalton Fen in what passes for Huntingdonshire, a moth-er called R.P. Demuth spotted one crawling on his 'lampsheet'. As was the habit of the era, he collected and pinned it. The specimen was bequeathed to London's Natural History Museum in 1995, where it hides spreadeagled among eight million other dead moths and butterflies.

But what of *our* female? For Matt, this is an unprecedented opportunity to learn more about the species and, if she has already mated, to gather eggs to raise in captivity then stock a new colony. With our endorsement, Matt elects to takes the female home. A few days later, once she has laid eggs, he will release her. When he does so, she will beetle across his palm before flying four metres into the herbage – proof that females fly. Matt will tender to the twenty-five caterpillars the female produces and release them into the wild. From the Holy Grail, hope.

The final hour of the night passes in a daze. We count ten males in total, only one of which actually makes it into the trap – the others simply sitting on the ground, awaiting discovery. But none eclipse the female moth. At 3 am, somewhere between night and dawn, skylarks begin to chorus. We pack up, shake hands, slap one another's backs in disbelief and depart.

Lincolnshire's Marsh Moths may be studied only a few times a year, but at least they are surveyed. Many potentially valuable sites are not accorded this privilege. Within its county reach, the Norfolk Moth Survey aims to make good that oversight. Led by walrus-moustached Ken Saul, this cadre of volunteer moth-ers focuses its ministrations at poorly known sites. At June's opening, they target the Norfolk Wildlife Trust reserve of Upton Fen, on Broadland's western frontier. Nobody knows the last time anybody ran a moth trap here, if ever.

We are late to join the gathering. Will and I sweat through the evening mug for half a mile into the central fen, carrying a weighty generator. Wellies trampolining along the floating meadow path, our passage is shaded from the reclining sun by the heavy murk of Alder carr. We are too late to admire the meadow's riches but it looks pleased with itself, showing off the candy-pink paper cuts of Ragged Robin, the foppish flamboyance of Yellow Flag Iris and the willowy enigma that are the Broads' Southern Marsh Orchids. A Grasshopper Warbler's song comes from a secluded reedbed as we unreel cables and position traps between brimming ditch and squelching, scruffy marsh. By the time we have powered up, a beer-bellied Woodcock is piercing the air in territorial dusk flight. It's time to join the others.

Since the Society of Aurelians formed during the first half of the eighteenth century, moth-ers have routinely gathered to share their interest and pursue new ground. Communal field trips were common, as were meetings in drinking houses and 'at home' sessions offered by grandees. Unlike birdwatching, say, the pastime is typically collective – a community of the like-minded and inclusive, welcoming of newcomers and generous towards beginners. Tonight, the fifteen of us rotate sociably between people's traps, gazing below the Perspex as entranced as mesmerised campers before a fire. We practise camaraderie rather than competition, and people variously solicit, query, challenge and divulge.

Tall and approachable, Stewart Wright proffers a pot that contains the crinkly flowerhead of a sedge, on which I discern a black, ant-like form. Looking closely, I realise that it is seven millimetres worth of micromoth. Shimmering gold and crimson wings lead to a fluffy black head, huge eyes and what looks like a beak. 'This is *Micropterix mansuetella*,' Stewart says. Now the 'beak' makes sense. The genus *Micropterix* comprises vastly primitive moths. These living fossils have changed little across 200 million years. Unlike every other British genus, adults possess functional jaws – all the better for nibbling solid repast, in this case sedge pollen. Sadly, mandibles are also cumbersome to transport, hence their disappearance from the evolutionary arms race.

Stewart openly admits to an obsession with small moths – the tinier and more obscure the better. He feels impelled to fill the gaps in biological knowledge, considering it a matter of fairness. 'I don't like things getting neglected. On fungi forays, I take the same approach, focusing on microfungi, which are chronically under-recorded.'

Encircling the trap, we share tales of how we came to moths. Broadland local Steve Smith recalls: 'My interest kicked off when a friend and I went to a moth evening in a churchyard. That inspired us to build our own trap. Things have escalated since.' His partner, Dot Machin, nods at this. Inhaling a clove-scented cigarette, the parish-council chair admits to long being scared of moths but says familiarity helped overcome her fear so that she now enjoys looking at moths when there are no birds to watch. 'And it's just wonderful being outside at night,' she exhales.

Over the hours we spend companionably with darkness, the star catch is a Dentated Pug, a jagged, carpet-like marsh dweller with few remaining strongholds in England. I am engrossed by various species of china-mark, whose larvae defy the moth paradigm by growing underwater, feeding on aquatic vegetation. One that we see, Small China-mark, breathes through thread-like gills running along its flanks. It is half-moth, half-fish. Somehow, adults emerge from their pupal lair while still underwater, struggling upwards to reach the air.

I marvel at a Lobster Moth luxuriating on the flank of one trap. Hairy legs stretch languorously, hindwings thrust forwards and outwards beyond the forewings, professing the composite moth to be nothing but a cluster of obsolete leaves. Its caterpillar is more bizarre still, a grotesque parody of a crustacean, combining protective armour with mimicry. Ulisse Aldrovandi, an Italian whom Carl Linnaeus described as the father of natural history study, was so bewildered by the beast that he thought it might be a spider. In *A Moth-Hunter's Gossip*, P.B.M. Allan considered the larva 'a crawling monstrosity … too grotesque to be interesting', but acknowledged that he might be alone in harbouring such disdain, explaining that Victorian entomologists so venerated Lobster Moth that they willingly paid £5 to procure a specimen – about £1,400 in today's money.

Come 2 am the rate of acquisition slackens, and Ken decrees the night closed. When he combines our lists, he tallies more than 150 species: no mean feat. But I know nothing of such success while heaving generators and traps back to the car, pockets clinking with glass moth pots. I finally slam the boot shut at 3 am, just as a muted glimmer christens the new day. I need my bed.

Two nights later, the moon still stands proud in the sky when I reach Sutton Fen shortly after 4 am, barely five hours since returning home from a trip to the south coast. As I pull up beside a barn that serves as workhouse and conservation office, a Hedgehog clockwork-toys across the track ahead, startling Mick A'Court, who says he hasn't seen one 'for ages'. Mick is assistant warden at this under-the-radar RSPB reserve. Although one of the charity's biologically richest holdings, its landscape is too sensitive to withstand public access. I owe today's visit to the kindness of this wild-bearded, fruity-mouthed Londoner. Like Josie Hewitt, Mick got into mothing through bird-ringing, twenty years previously. We are joined by Lynnette Nicholson, a moth-er who ascribes her interest to noticing day-flying moths while out birdwatching. Shortly after fleeing the capital for Norfolk's quieter life, she received a moth-trap for her birthday, which appealed to her innate curiosity. 'I like to know what things are,' she says, 'to identify them, give them a name.' She now religiously runs two traps in her village garden.

We sluice through the dew-dropped fen, carving across the reedbed-edge territory of Sedge Warbler and Reed Bunting. Distant and unseen, a Common Crane pair bugles greetings – their reedy, oboe-like calls redolent of wild openness. 'I watched changeovers at the nest,' Mick says of this rare but resurging British bird, 'but nothing hatched. I think this new set of callers are young birds going through the motions.'

Cranes are a bonus. Lynnette and I have joined Mick in the hope that his six traps, set overnight in places known by reserve staff as 'Bumpy Bank' and 'The Motorway', have seduced the reserve's Broadland moth specialities. We are racing against the sun, rushing to release moths from the traps before they overheat. As Mick checks the egg-trays, I transcribe his mutterings. Mick is adamant that I must note down absolutely every moth that he shouts out. 'I count everything,' he emphasises. 'I don't go round just ticking things off. I like datasets from regular sites.' That way you can run analyses, he explains, determine patterns, work out whether something is doing well or poorly. This morning, he is slightly downcast by both numbers and diversity – the impact of the previous night's full moon, he reckons. Nevertheless, we effortlessly surpass a hundred species – and the catch oozes quality.

First up are numerous, freshly emerged Drinkers. They are fat-lad moths with a Womble-like 'snout', huge vacant eyes and ridiculously broad, leaf-like wings. The female is massive and golden; the male is smaller and hewn from brick. The moth is named after the proclivity of its gargantuan, hairy caterpillar to sup raindrops and dew – a habit first described by Dutch naturalist Jan Goedart in 1662. 'Seeing that Drinker takes me back to when I started mothing,' Lynnette murmurs. 'When you catch your first Drinker, you simply can't believe it is real. It makes you question what's out there that you don't yet know.'

Water Ermine makes an early bid to steal the show. A handful of black poppy-seeds spot long wings of driven snow, atop which sprawls a judge's ermine. A scarce species, its entire British range hugs the coast between Norfolk and Sussex. This is the first Lynnette has seen. Similarly sparsely marked is Dotted Footman, a Broadland exclusive. It is silkily grey, almost pear-shaped. This individual is Mick's first of the summer and another new species for Lynnette. Its appearance prompts me to muse whether we might also catch the closely related Small Dotted Footman, an even more restricted Broadland critter that eluded discovery in Britain until 1961. Mick argues that it is too early, and he is proved right. I must wait another six weeks before seeing it this year, at the RSPB's Strumpshaw Fen. It transpires to be an intriguing summer for this nationally near-threatened species, which appears at several novel sites, prompting suggestions of range expansion.

A third near-threatened moth, Reed Leopard, inhabits only three areas of Britain, notably Cambridgeshire's Fens and Norfolk's Broads. But its scarcity is surpassed by physical sexiness. The colour and texture of an age-dried reed stem, this moth wears a furry shawl and sprawls forelegs as might its namesake feline. Its peculiarly rounded wings retreat backwards interminably, yet a preposterously lengthy abdomen somehow still protrudes beyond them, sticking out like a thumb-tip beyond a plaster. 'How can you have a body like that?' Mick wonders. 'I mean, what's the point?' Unsurprisingly, Reed Leopards are weak flyers, struggling as much with the weight of their abdomen as I do with a generator. We end up catching twenty-one. 'So many!' Lynnette coos. 'Oh, this is nothing,' Mick gruffs. 'That's a low count.' After seven years spoilt by Sutton Fen's riches, he admits to having become blasé. 'Trouble is, these things are common here. You expect them. You don't think about them any more – oh, just another Reed Leopard. But they're bloody rare elsewhere, I know that.'

More than almost any other Broadland moth, Reed Leopards have *cynefin*, which means 'haunt', 'usual abode' or 'habitat' in Welsh. Reed Leopard is a moth of its place, both geographically and ecologically. It doesn't stray beyond its exacting context, making it an ideal indicator of environmental health. It refuses even to grace Strumpshaw Fen, just 8 miles south-west. This is a risky life strategy. Within the Ant Valley, Sutton Fen merges seamlessly into Catfield Fen, Norfolk's most controversial and possibly most endangered reserve. Since 1986, farmers have abstracted water adjacent to Catfield Fen to irrigate arable crops. Concern was first raised in the 1990s that this agricultural water-removal was parching the Fen. Since 2012, RSPB surveys have confirmed the suspicion and revealed that parts of the site were becoming more acidic. Catfield's Swallowtail butterflies, rare Fen Orchids and scarce moths – including Reed Leopard – were in danger. In 2015, following advice from Natural England, the Environment Agency refused to renew the farmer's water-abstraction licences. The irate agriculturalist appealed, pushing the case to a public inquiry. In 2016, the inquiry inspector concluded there was no justification to accept the appeals 'when weighed against the conservation interests that would be harmed'.

For nearly three years since, Catfield Fen – and Sutton Fen, by association – have seemed safe. Catfield has started its road to recovery. A few weeks before my visit, however, the *Sunday Times* ran a story suggesting that 'Britain's first water war [had] broken out' – right here. In order to preserve the Ant Valley reserves' water needs in the context of a warming climate and growing human population, the Environment Agency had suggested that more than twenty water-abstraction 'licences could potentially be removed … including two of Anglian Water's licences to abstract water for public supply'. Broadland wildlife held a jubilant party, but local politicians were up in arms that villagers' needs played second fiddle to the environment. Even with subsequent changes to Anglian Water infrastructure, Catfield Fen looks set to remain an indefinite battleground between the demands of the environment, agriculture and human ablutions.

Our Fenland forays continue through June. Malcolm Hillier invites us to spend a night in a brownfield site on the outskirts of Peterborough. Swaddywell Pit has a harlequin history – medieval quarry, landfill site, racetrack, illicit rave venue – but is now managed as a community nature reserve. Here we catch up with Concolorous, a less-boring-than-it-sounds moth forced out of Fenland by drainage that has founded a vibrant domicile within the grassy, scrubby, flooded quarry. Our paths again cross with Malcolm at a sunny-morning event in Fenland proper, at the suggestion of conservationist James Symonds. Nourished by James's unbridled enthusiasm, we join a naturalists' ramble through a calcareous, herb-rich meadow at Chippenham Fen. This precious, isolated remnant of Fenland sits at the southern edge of the once-saturated region. Chippenham serves as one of perhaps only four remaining British locations for Silver Barred. This sumptuous day-flying moth – all carrot-cake wings, drizzled with icing – once flew commonly across the entire Fenland region but now finds only two homes here. We see a score, each encounter thrilling yet saddening, mounting jubilation tempered by a deepening apprehension of our forefathers' environmental impact.

Concolorous and Silver Barred excite, of course. But as with so many of this year's ventures, the anticipation involves known unknowns. We know our target moth lives here, so it will not be a revelation to see it; the tenterhooks emanate from the unknowns of whether it has emerged and whether it will enter our traps or nets, in whatever conditions greet us on the precise day we visit. Nevertheless, we travel with a sense of, if not entitlement, then assuredly expectation.

There is none of that with Marsh Carpet. This is a moth of unknown unknowns. Once uniting East Anglia's two great damp regions – Fenland and the Norfolk Broads – this toffee-coloured dreamboat disappeared entirely in the 1940s, and was feared extinct. Marsh Carpet remains fickle to the point of wholesale unpredictability, thriving at a site briefly before disappearing, inexplicably and irrevocably. It has previously established colonies far from its strongholds, in Yorkshire, but those now appear to be consigned to the memory bank of what once was where. Similarly, it has disappeared from two key Cambridgeshire reserves.

Norfolk presumably remains a stronghold, but the first decade of this century elicited just five records. Suffolk is likewise a key county, but Marsh Carpet has gone missing during the past three summers. Keeping tabs on adults is near impossible because they display little interest in moth-trap lights. When I ask moth-ers where we should try and see it, they shrug. Nobody knows.

And yet. This quagmire of uncertainty provides opportunities. To aspire to discover a new domain for Marsh Carpet. To help Butterfly Conservation's Sharon Hearle as she launches a project to unravel the shy moth's secrets, the better to save it. To contribute. And to do so as close to home as possible.

So I compile recent records of Marsh Carpet from garden moth-traps around Norwich – I have seen two of them, but know of several more, including one barely five minutes' stroll from my own. Triangulating the catches, I pinpoint a promising stretch of damp valley meadows alongside the River Wensum in the north-west of the city, less than a mile from home. The trail leads to a disused fish farm managed by the Environment Agency. Might this conceivably be a hitherto undiscovered Marsh Carpet breeding site?

Dougal McNeill, a friend working at Natural England, yields the necessary introductions. It turns out his colleague, Adrian Gardiner, had already conjectured that Marsh Carpet might occur at the fish farm, now used as a storage depot. Even better, he had mapped a patch of the moth's foodplant, Common Meadow-rue, alongside the Agency buildings. The Agency's site manager, Steve Lane, transpires to be a closet conservationist and eagerly agrees to let us search for Marsh Carpet. Taste buds already whetted by having attended a moth event, Steve expresses keenness to accompany us. 'This place has been part of my life for twenty-five years now,' he says. 'Alongside David Attenborough, it awakened my interest in the natural world.'

The stars may be aligned, but there is no chance of this whimsical moth actually being there, of course. The world doesn't work like that.

And yet. After a chilly, windy afternoon, the first Saturday evening of July becomes mild and calm, muggy even. It's a night for an impromptu tipsy barbeque or beering in a pub garden. Shunning such bacchanalia, Steve, Adrian and I meet by a gushing weir that eels ascend when their time is right. After much harrumphing over whether an early night might not be a better use of his depleted energy budget, a drained Will huffs up. As Steve unlocks an imposing gate, friend Dave Andrews puffs an approach, sprinting from his nearby home after finally coaxing his toddler to sleep.

We bump along a track that divides the rivers Wensum and Tud until we reach the Environment Agency outbuildings. Steve at our head, we caterpillar along pathways of sunset gold sitting pretty above the old fish pools, their rectangular forms now thronged with sedge, reed and other Fenland greenery ignorant of the concept of personal space. After years of disuse, botany has regained its rightful berth. Perhaps too much so. 'The herb fen has expanded in the ponds since I was last here,' Adrian observes. 'Most of our sixteen ponds have regressed,' Steve agrees. 'We've lost open water, which is sad given this place has had eighteen types of dragonfly – about as many as any site can get.' Until 1993, the

National Rivers Authority used to farm fish here to restock angling sites. For a 'halcyon five-year period after that, we had a proper management regime,' Steve continues. 'There used to be Ragged Robin and Southern Marsh Orchids everywhere. One day I saw Bittern and two Otter.' Then the money ran out.

Nevertheless, Will, Dave and I like what we see. Steve rightly calls the place 'a little gem'. Most excitingly, two decent stands of Common Meadow-rue are in full stately bloom, frothing butter-yellow flowers. Tacitly and improperly, we allow ourselves to dream.

As dusk whispers an entrance, the air above the pond-sedge starts to dance with the wan forms of wainscots – tentatively at first, as at a school ball, then vibrantly, shamelessly, incessantly. It is a good omen. So too are Dotted Fan-foots – too many to count – flattening their brittle, equilateral-triangle forms on the pinnate leaves of Meadow-rue or twiddling through the air between plants. Round-winged Muslin are airborne in similar abundance, our swishing nets repeatedly folding around their soft-shaped forms. We swipe two nondescript but nationally scarce micromoths, both new to me: *Pseudopostega crepusculella* and *Thiotricha subocellea*. This is most positive. We crank up the generators, throwing white into the black.

About half-ten, Will pops to his car for some pots. As he saunters back, his eyes catch a moth silhouetted on the illuminated window of the building where we supped a pre-dusk cuppa. His befuddled brain takes time to compute. But then he clicks: hugging the pane is a Marsh Carpet.

Will tiptoes forward, fumbling for a pot. Marsh Carpets are only weakly attracted to light at the best of times. They are skittish creatures too. This moth could take flight at the slightest jerkiness in his approach. But the Wingman does it. He pots the moth, gently but securely. Then hollers into the night: 'Marsh Carpet! I've got one! Marsh Carpet!'

I am orbiting the outermost traps we have set. Pivoting immediately, I hurtle towards the depot. We assemble, me breathless, everyone stunned and yelping for joy. The moth is glorious – a gently curved piece of fudge through which run two thick veins of liquorice, each dusted with sugar. Against all conceivable odds, we have found a wholly new site for one of Britain's rarest and most poorly understood moths. Across the valley, in the housing estate, some thoughtful soul salutes our success with fireworks.

Stoked, we continue around the traps. Twenty minutes later, Steve bumbles up and mentions that he's photographed an odd moth on the depot's rear window. He pulls out his phone to show us. It's a second Marsh Carpet. We race to the building. The moth is still there, still real. Despite jitterbug hands, I pot it. One Marsh Carpet is remarkable. But two?

And yet. At half-eleven, Adrian peers into his trap – the one sited beside the Meadow-rue. He can't believe his eyes. There, sitting pretty, in flagrant abandonment of the disdain its species traditionally holds for MV lights, is the third Marsh Carpet of the night.

If one Marsh Carpet is remarkable, what are three? As far as we are aware, no Norfolk site has ever had more than two adult Marsh Carpets on a single night. Not only have we discovered a new breeding colony, but it appears to be healthy too. We are incredulous, elated by the implications, excited by the possibilities.

In the coming weeks, I put Steve in touch with Sharon Hearle. Over the months that follow, they meet at what Sharon decrees 'an intriguing little oasis, hidden away' and agree a conservation action plan designed to benefit Marsh Carpet and to understand its mysterious ways. 'We're really keen to do what we can conservation-wise on the site,' Steve later tells me.

In a resurgent, urban-fringe fen just west of the Norfolk Broads, far from tourist honeypots, there is the sense that the moth of unknown unknowns may be about to become knowable.

10
Sylvan secrets

Dorset, Berkshire/Oxfordshire, Kent, Hampshire and Cumbria
May and July–August

More than any other landscape, mothing in forests demands nocturnal exploration. Only during the inky hours do woodlands reveal their innermost secrets, creatures that shun daylight every bit as much as a vampire. Among trees, the otherworld of night is when the wildest things appear. This timetabling presents problems. Humans' senses are relatively ill-equipped for life without light, and my own natural rhythm involves waking at dawn and yawning by dusk. Woodland mothing requires inverting long-engrained circadian habits.

Intense activity from dusk till dawn also demands embracing our fear of the dark. Before we learn to synthesise what limited night-time data our senses are competent to furnish – the smell of damp earth, the airiness of a glade, the obfuscation that signals an impenetrable thicket – our imagination plays conjuror, extrapolating and interpreting erroneously to feed our insecurities. Trunks become shadows become monsters. Fallen branches transform into poised serpents; mist morphs into phantoms.

Yet it is at night in this sylvan realm that I learn to feel most alive. To find myself, as Roger Deakin wrote in

Wildwood, by first becoming lost. The rupture from chaotic modernity – all blue-light screens and deadlines and overdrafts – engenders calm. The solitude no longer disconcerts but stabilises. I dredge up a recollection of a nineteenth-century verse by the German poet Heinrich Heine. The title – '*Waldeinsamkeit*' – is untranslatable but essentially means to be both alone in and at one with (*Einsamkeit*) the forest (*Wald*). The literal lonesomeness is imbued with calm, with belonging.

It is time to venture into Britain's wildwoods.

The north Dorset woodland goes to sleep about 9 pm. The odd Robin *tick*s as a solid breeze rustles the greenery of slender Beeches, but that's about it. Distant motorbikes hint at an outside existence, but our present world is one of thickening air, the aqueous light of a kelp forest and dense, dark Yews that germinated under the rule of William of Orange.

The north Dorset woodland also comes alive about 9 pm. Moths unseen by day, lying flat against the underside of leaves, stir. Phantom-white in the murk, they wisp upwards like reverse confetti. It is time to clock on for our nightshift.

It is mid-May. The Wingman and I are helping Butterfly Conservation's Fiona Haynes survey a moth whose UK recovery is being marshalled by the Back from the Brink conservation programme. Barberry Carpet is the sole troubled moth to warrant its own project. That alone speaks volumes about its worrisome status, officially considered endangered. Whereas

Barberry Carpet once ranged north to Yorkshire, it now subsists in just twelve locations in four counties. The Dorset population – at a location withheld from the public – may be the country's most robust.

As Barberry Carpet is one of eight strictly protected moths, only licence-holders such as Fiona – and those working with her, such as us tonight – can legitimately search for it. Aided by a quintet of Dorset-based friends, Durwyn Liley and Phil Saunders among them, Will and I place our traps at Fiona's service. She welcomes us by clarifying the Barberry Carpet's existential problem: essentially, the moth is collateral damage. The moth's caterpillars feed solely on Common Barberry, a yellow-flowered shrub of woodland edge and hedgerows. The plant is an intermediate host for a wheat-stem rust fungus that once blighted cereal crops. Rust-resistant wheat varieties solved the issue but, perhaps as late as the 1990s, farmers would grub out Barberry growing near their fields. Without food, Barberry Carpet populations died out – a manifestation of the perils of strict dependency.

Fiona briefs us on the project's aim to re-establish the moth's habitat by planting three thousand Barberry bushes across fifty-odd sites spanning three counties. 'In Gloucestershire and Wiltshire, we want to connect existing populations. At this isolated Dorset site, we want moths to spread into the countryside,' she explains. A greater number of more resilient populations is critical because there are worrying hints that wheat-stem rust may return to Britain; a sighting in 2013 was the first for six decades. Fortunately, Fiona is working with crop scientists who advise that planting Barberry at least twenty metres from arable land is safe.

Hedgerows provide key habitat for many moths, of course. Douglas Boyes – the academic who favours an 'informal, open dialogue' about nicknames for micro-moths – estimated perhaps 21,000 moth caterpillars residing in a hundred-metre stretch of one 'uninspiring Hawthorn hedge' he surveyed. It follows that habitat management on agricultural sites can help Barberry Carpet. 'We're working with farmers to encourage cutting of hedges later in autumn, once the caterpillars have safely pupated underground, and to cut on rotation,' Fiona explains.

At Fiona's behest, we place light traps along a Barberry-flecked hedge fringing the forest and beside planted Barberry within the woodland. Fiona's decision to survey adult moths is a break from the traditional approach of counting caterpillars – and is also more likely to engage the public. 'When we first tried it last year, we had no clue whether it would be successful,' she says. 'We were over the moon to catch five adults.'

The woodland hushes, expectant. In my headtorch, dense bunches of a dozen globular primrose-toned flowers spray from the underside of woody Barberry branches that strain over the pathway from the main hedgerow. There is foodplant enough, but is there moth? We learn soon enough. Within ten minutes of the 9 pm nightshift clocking on, Phil spots an adult. He calls excitedly. We gather rapidly.

Barberry Carpet has the typical open-winged posture of the family Geometridae, its forewing edges set at a mildly obtuse angle and meeting along a gently concave trailing edge. It is smartly patterned, with wavy stripes of cocoa, espresso and cappuccino. The rear half of the wings bears paler tones, as if a child got bored midway through colouring it in.

Flying low over the ground, hugging the hedge, further Barberry Carpets are soon found. Phil breaks new ground when watching one nectar on a Barberry flower. It has long been known that caterpillars subsist on Barberry foliage, but Fiona is excited by the revelation that the plant also sustains adults. 'That Barberry is a nectar source further strengthens the argument for planting it,' she buzzes.

By 10.30 pm, I seem to be the only surveyor not to have found the special moth. Finally I spot one and close in with camera. At that very moment, my mobile phone rings. It is late for anyone to call, let alone my father. This suggests bad news.

My dad is crying. My dad doesn't cry. This means the worst kind of news.

At the precise moment I learn that cancer has finally overcome my cherished uncle Colin, the Barberry Carpet seizes its opportunity. It flicks its wings in anticipation, then flees upwards and outwards and onwards. Recalling the old belief that moths represent human souls departing the body, I watch the Barberry Carpet depart as I bid Colin farewell and weep.

Fifteen minutes later, guilt challenges grief. My phone rings again. This time it is the mother of the family where my daughter is having an enforced sleepover as a result of my absence and my wife Sharon's current incapacitating bout of sciatica. Having been unable to rouse my painkiller-stupefied spouse, Claire has resorted to calling the truant dad. Claire wouldn't ring unless absolutely necessary; Maya must be ill, injured or upset.

She hands the phone to Maya. Two hours after lights out, my eight-year-old is awake and hysterical, glugging through tears. 'I can't get to sleep.' Gulp. 'I miss you, Dada.' Sob. 'Why have you gone away?' Wail. 'Why have you left me?'

As the consequences of this selfish quest gouge my heart, I try to talk Maya down. Hugging her tight through words, I smother her insecurities and unhappiness with love. Eventually, she stills. Everything is going to be OK. Wept dry, she returns the phone to Claire. I thank her and hang up. I try to reorient my brain, strive to resume the search.

Two hours later, we have counted a mind-boggling fourteen Barberry Carpets along the hedgerows. Checking our traps for the first time, we find they have been successful too. We total seventeen moths – surely more than anyone has caught in a night for years, and confirmation that searching for adults is a valid survey technique. Moreover, Fiona has a few moths to show at a public event the following morning before releasing them unharmed to continue their lives.

'If there were a Nobel Prize for Ecology, and you could award it to a place rather than a person, Wytham Woods would surely be a prime candidate.' Such is the judgement of Professor Lord John Krebs, former government chief scientist and doyen of ecologists, in his foreword to the book *Wytham Woods: Oxford's Ecological Laboratory*. Krebs argues that this 350-hectare woodland west of Oxford and owned by the University of Oxford since 1942 is 'almost certainly unmatched ... as a place of sustained, intensive ecological research extending over nearly three-quarters of a century'.

Krebs cut his academic teeth at Wytham Woods in the 1960s, an early participant in a seminal study of Great Tit ecology that continues today. Much of the material underpinning a book I wrote on Badgers a few years ago derived from long-running studies of the mammal's ecology in this same woodland. George Varley's study of Winter Moths here arguably engendered the whole concept of population dynamics. As part of Varley's heritage, we are focusing on moths tonight. If we can get in, that is.

With the sky bruising rapidly, Ben Lewis, Will and I are stuck on the wrong side of an unequivocally locked, formidably tall gate while our host, Douglas Boyes, encounters voicemail after voicemail of university folk who might conceivably know the new entry code. Eventually, he persuades a local resident to reveal the numbers. Open sesame.

In his early twenties, Douglas is a lepidopteran prodigy and flamboyant soul, his trademark attire starring a colourful shirt decorated with butterflies and moths. Alongside his academic research, Douglas is under contract to provide a baseline survey of Wytham's moths, which is why we are here. 'The Woods haven't been seriously surveyed since the 1980s,' he explains. 'I plan to run traps fourteen nights a year for three years. The updated moth list will then be used to springboard research – such as investigating how different ecological management techniques impact certain species.'

Tonight we have the varsity playground to ourselves. As we position twelve traps in the quiet, there is nobody else for miles around. We place lights under iron-barked Beeches, beside hulking, elephant-skinned oaks, and under the protection of Small-leaved Limes whose lowest boughs flirt wearily with the bare ground. But we do so ineptly. Ben's new generator fails at first start, smoking and stinking into rapid obsolescence while Will's traps, surreptitiously plugged into an external socket of a barn, blow the electrics.

Eventually, incompetence surmounted, we are up and running – the lights making lightning forks of branches above – just in time for wind-rattled skies to launch a storm scheduled to last three hours. Moths batten down the hatches and we retreat to the barn, where we gather like Macbeth's witches around the cauldron of a light trap. Will pours damson-gin homebrew into moth pots in our makeshift pub, as Douglas recalls his early mothing exploits in parts of mid-Wales where nobody had done any serious mothing: 'Discovering new sites for scarce species like Ashworth's Rustic was really rewarding.' He giggles and says, 'A policeman turned up once, having thought I was rustling sheep. It transpired he had been to a moth-trapping event and preferred moths to butterflies. We chatted for ages.'

To a backdrop of yowling Tawny Owl, conversation turns to his PhD thesis. He aims to help determine why so many moths are declining. 'I want to understand whether light pollution – artificial light at night, from streetlamps and the like – has a population-level impact on moths,' he explains, 'and, if so, to consider how serious this is in the context of climate change and habitat loss.' To find out, Douglas is comparing caterpillar abundance and diversity at well-lit rural sites like roundabouts with equivalent non-illuminated habitat. His initial desk-based research reveals evidence that unnatural light exerts diverse impacts across most life stages of moths and on key behaviours – from inhibiting feeding to reducing mating. This jolts me: might running a moth-trap harm the creatures we are studying and have come to love?

Douglas explains that the problem of light is widespread: much of Europe experiences nocturnal sky-brightness levels more than 50 per cent above the natural. Modern LED lighting is likely to be more harmful to moths than traditional sodium streetlights, so the problem may worsen as the technology spreads. Nevertheless, to date, Douglas has unveiled reassuringly 'little direct evidence that light pollution has negatively affected moth *populations*'. Although streetlighting and my moth traps might interrupt individual moths' going about for a few hours, neither should exert substantive impacts. This accords with analysis of findings from the long-running Rothamsted Insect Survey. Across this nationwide network, the biggest changes in moth populations came from traps in grassland and woodland – habitats unlikely to be greatly affected by light pollution.

Still, there remains much more to learn – and that means lots of real-life research. Douglas's work involves sweeping grassy verges and beating hedgerows to reveal caterpillars – activities that will end up occupying Douglas for 329 hours over twenty-eight days and thirty-three nights during his first year of fieldwork.

But enough pub chat. We retire for three hours, rolling sleeping bags onto the open-sided barn's hard floor. I try not to think of my family, warm and comfortable, in their own beds.

Dawn follows midnight with indecent haste, and we scramble to empty traps before dog-walkers arrive. After the unpleasant weather, our catch is understandably inauspicious. Under 300 individuals of fifty-one species is a paltry offering from a dozen lights.

The night's saving grace sits on the outside of the final trap: an unfamiliar pug. Comprising a baffling multitude of predominantly grey and unremarkably patterned species, pugs are an acquired taste among moth-ers. But this example is both subtly attractive – with waves rather like a comforting wool jumper – and mysterious of identity. Its name doesn't trouble Douglas, fortunately. Excitedly, he pronounces it a Fletcher's Pug – famed for being one of just four larger British moth species that Victorian lepidopterists overlooked. Even today it seems hard to come by. A rare, lime-loving species, this is not only the first for Wytham Woods, but only the sixteenth for the county of Berkshire (to which the site's moths are anachronistically allocated). Coming at the start of his three-year survey, Douglas is thrilled. It is the first of fifty newbies that he will find here later this year – hitherto-undiscovered secrets revealed by the famous forest once Douglas has earned its trust, pushing the site's moth list above a thousand.

'The popular locality known as "Hamstreet" refers to Orlestone Woods,' Kent moth-er Michael O'Keefe wrote in *Atropos* magazine, 'and constitutes what is generally considered the finest locality for woodland moths in Britain'. Orlestone comprises mixed, damp woodland atop weald clay; its tree-life is particularly marked by oaks, Hornbeam and Aspen. Seemingly overlooked by Victorian entomologists, its potential was recognised by local moth-ers during the 1930s. Partly thanks to easy railway access, it rapidly came to national prominence for its array of rare breeders found nowhere else in Britain.

Sadly three lepidopteran stars of Orlestone's post-war heyday – Lesser Belle, Lunar Double-stripe and Clifden Nonpareil – soon vanished as British breeders. The Clifden has seemingly returned in very recent years, but its berth among the moths of the disappeared may already have been nabbed by Feathered Beauty, for which local moth-er Bernard Boothroyd knows of no record since 2010. They may only ever have been temporary colonists that, suggested E.B. Ford in *Moths*, 'adjusted themselves' to these particular woodlands, for the habitat was not unlike that found on the adjacent Continent.

The moth fauna that survives today nevertheless justifies O'Keefe's accolade. Accordingly, Will and I plot an unseemly number of nights trapping here this summer. We devote the first brace to Longrope, part of the complex managed by Forestry England. Here the erupting limbs of oaks, Beech, birch and Hornbeam are neighbourly, and Meadowsweet leads us past a cemetery of blackened tree stumps. Five local moth-ers join us, plus James Hunter who arrives with a car full of traps. James has done five straight nights' mothing on his own during a week off work, but is now craving company on lepidopteran missions.

We position nine traps in clearings and along broad grassy tracks. 'Put one right under the old oaks if you're after Scarce Murvell,' James advises. He is referring to Scarce Merveille du Jour, the night's principal target. Having seen this moth in Spain, I know it to be exquisite – a vibrant mint-green creature whose black shadows meld it effortlessly with verdant foliose lichen that wraps around branches. I yearn to see it in Britain, but its genus name – *Moma*, from Momus, the Greek god of mockery – suggests Mother Nature may scorn us tonight.

Even so, the overcast, muggy evening starts promisingly. Netting flying moths as the light fails, we encounter an Orange Moth. The large, butterfly-like tangerine flops along trackside herbage. As we wait for darkness, local moth-er Nick Green recounts how he discovered the importance for wildlife of a nearby woodland, Alex Farm, and later came to buy it. The saga involved erecting gates to deter off-roaders and pheasant-shooters and persuading the previous owner to change his will. Nick and his wife Sian now run the farm with wildlife in mind.

It is a heartening conversation, but even tales of hope must fall silent when moths start flying. And how they fly. The forest is alive. It is the most intense night of mothing I have ever experienced. We are assaulted by chaos. By dawn, we have trapped 3,000 individuals. Probably the same number again are attracted to the lights but never even reach the trap, instead blizzarding around our face, alighting on our shirts or littering white groundsheets beneath our traps. We identify just shy of 200 species. Every trap round elicits something noteworthy. Pockets clink with the success of moth-filled pots. It is hard to keep up.

The first star is a sumptuous micromoth, *Agrotera nemoralis*, of which we catch at least thirty. A golden cloak behind its head flares into chestnut-velvet wings that terminate in a zebra-crossing trim. Twenty years ago, those running a scheme recording rare micromoths lamented in *Atropos* magazine that they had 'received no record of this species for the last five years, and news of its continued presence in east Kent or elsewhere would be most welcome.' Since then the fortunes of this Hornbeam specialist have sky-rocketed. 'It wasn't in Orlestone until five years ago,' Bernard Boothroyd explains when we catch the first. 'Since then it's gone bananas. This year I've had seven just in my garden.'

My attention is successively pinched by moths that I have seen only occasionally in previous years. Several Large Emeralds are swoonsome – enormous, luminescent Kermit-green creatures that outclass any British butterfly. A smaller version is Blotched Emerald, whose deftly placed tan-coloured blemishes kid would-be predators that it is merely a decaying leaf. There are a hundredfold more Lackeys than have previously crossed my path, many lovemaking. This moth is named in recognition of livery lace worn by both eighteenth-century servants and, with a modicum of imagination, its caterpillar. The adults look less distinguished – stocky, hairy creatures, ribboned in beige and brown. Dozens of Festoons, a specialist of southern oak woodlands, are diminutive chestnut triangles that erect their abdomen tip between drooped wings. At least one Great Oak Beauty, a near-mythical moth hitherto thought absent from Orlestone, is 'the size of a light aircraft,' Will reckons.

But where are the Scarce Merveilles? 'They come in late,' James says. 'Be patient.' As ever, my inaugural moth mentor is on the money. From 2.30 am onwards, we find five around our traps, all – as he predicted – beneath thickset, rivulet-barked oaks. These peppermint sweets are every bit as lovely as intimated by their vernacular name, the French for 'wonder of the day'. They provide the night's climax – and its close. As the sky lightens with eastern promise, a female Ghost Moth grants herself a final flight, bombing low over vegetation to lay eggs while a Nightingale initiates an otherwise perfunctory dawn chorus. We reluctantly pack up, then make for bed.

Throughout the following day's come-down, Will and I chomp at the bit to return at dusk. But, although the location is the same, everything else differs. We are alone for a start, sitting in our camping chairs, ears cocked at undergrowth rustles. Clear skies and an onshore breeze render the night contrastingly cold. This means there are almost no moths, and our catch plummets by 90 per cent, with barely half the diversity. It is a salutatory lesson about the fickleness of mothing and the primacy of weather.

But it's not all bad. Checking Will's actinic, I spy a thumb-sized moth sprawled along the power cable. My heart leaps. Ever since I first flicked through a field guide, I have wanted to catch a Goat Moth. Now I have. Or Will has. Same thing, I tell him.

A rare resident in Britain, this primitive moth is a strange-looking soul. Large eyes dominate a tiny head swamped by a swollen thorax leading into bulky wings that seem caringly engraved. The moth's caterpillar is weirder still, emitting an ungulate-like pong that justifies its name and living inside a tree trunk for up to five years until it has digested enough cellulose to warrant pupation. Simon Curson, a Natural England ecologist, recalls a bewildering scent like fermenting apples issuing from three oak trees behind the shed in his suburban garden. One day, when he was slaving away in service of wildlife, a text from his wife provided the first clue. 'Mandy said there was an odd giant caterpillar shuffling across our lawn,' he remembers. 'Fortunately she had potted it and sent me a photo.' As it was half the size of a hand and vivid red, Simon quickly identified it as a young Goat Moth. Reckoning the larva was unhappy being imprisoned, Mandy told Simon she would release it. To bemused looks from colleagues, he frantically beseeched her to keep the caterpillar until he returned home. 'No, do not release the caterpillar!' he cried. Once home and with the larva safely admired, Simon discovered holes in trees behind the garden. 'I ripped off some bark and found five Goat Moth cats living there,' he remembers. The larvae had so damaged one trunk that the tree had to be felled.

Given the rapidly chilling night, we reckon that Goat Moth is as good as it will get, so call it quits and drive back to our digs as the sky starts to seep with eastern promise.

Tentatively, I bump the car along a rutted forest track while Will feigns deafness to the chalk-on-blackboard scratching of bramble thorns through paintwork. Reaching an open space, we swing around to prepare for the arrival of Kent moth-er Ian Roberts, who showed us Dew Moth two months earlier.

July is always busy for Ian. Every couple of nights, he runs traps throughout the dark hours, grabbing three or four hours' kip before heading into work. 'I struggle in afternoon meetings,' he admits, 'but you have to do it while the moths are flying.' He has given up telling his wife he will be back early – because he knows he won't. 'Mind you, if she's asleep when I make it home, I knock a couple of hours off my return time.' I prepare for a long night.

Joined by West Midland moth-er Darren Taylor, who is bunking off a holiday nearby with his girlfriend, and Norwich naturalist Dave Andrews, we are trapping at a secret woodland for a secret species. Well, almost secret. In 2018, Ian announced in *Atropos* magazine that Dusky Hook-tip had colonised a woodland near Saltwood in east Kent. Previously, this European species was known only from a hundred migrant examples. Ian and his friend Brian Harper had discovered its clandestine residency. 'We've trapped here regularly for a few years,' Ian recalls. 'One night in May 2017, almost the first moth to arrive was a pristine Dusky Hook-tip. In total we caught seven that night. Because conditions weren't right for migration, this suggested that the species had colonised.' To test the theory he returned in late summer to see if there was a second generation. There was. 'We caught thirty!' Britain had a new breeding moth.

We spread traps three cricket strips apart along a public footpath. The woodland is scrappy, but generous oak arches spray high overhead. There are also slender Alders and birch. The leaves of all three trees nourish Dusky Hook-tip caterpillars.

To complement the lights, I use a technique from the world of Victorian moth-ers: 'sugaring'. Two brothers from Essex – Edward and Henry Doubleday – are credited with discovering that daubing trees with a blend of treacle, brown sugar and alcohol attracted moths. In 1832, Edward published his findings, changing moth-ers' approach forever. It was molasses for the masses. Some Victorian moth-ers even pinned their calling cards to favoured sugaring trees, claiming territorial rights over potential usurpers.

Across some low branches, I fling lengths of rope simmered in Spanish wine, Adnams' beer, treacle and mushy bananas – plus a splash of suspiciously cheap rum. In the absence of transportable lights, offering a sugar rush was the go-to method for moth-ers until the mid-twentieth century. Even today, it remains an efficient way to see particular families, notably Noctuidae and Erebidae.

One member of each grouping swiftly appears at our ropes. A large noctuid with broad, dark wings is first to imbibe. The Old Lady – it really is called that – unfurls a long, curled proboscis and inserts it in the mixture. Moths have been doing this for 250 million years – curiously, a past more distant than that inhabited by the flowering plants with which the tongue-like structure had been assumed to co-evolve. A Bloxworth Snout – an intricately patterned triangle with protuberant palps – joins the bar. This moth has brethren with wackier tastes: some South American erebids lap nutrients from the tears of roosting birds, while a Malaysian cousin is a vampire, able to pierce skin with its proboscis and suck up the blood.

This pair of moths is exciting. But our gambit nurses a loftier ambition: to seduce a Dark Crimson Underwing into a spot of carousing. The size of an Old Lady but with flaming scarlet hindwings, this rare cousin of the

Red Underwing is barnstorming through this summer. During 'an influx more substantial than any I can recall,' veteran moth-er Paul Waring writes later in *British Wildlife* magazine, ample coastal migrants have even reached Northumberland. Intriguingly, suggestions are rife that this sought-after moth has reclaimed historical breeding woodlands in Kent to complement long-held residencies around the New Forest. If that is true, we might discover a new breeding secret tonight.

Accordingly, while we wait in hope for Dusky Hook-tip, I repeatedly check the ropes. At one, I jump at the sight of a large, dark moth with red hindwings slurping away. It starts at my clumsy approach, wafting away. Teasing my ineptly swished net, and flashing red undergarments to disorientate us, it zigzags stubbornly upwards and vanishes. We are convinced it was our quarry but, without confirmatory evidence, can claim nothing.

As frustration stumbles towards despondency, Dave manages to salvage our spirits. Returning from a trap round, he passes me a pot containing something vaguely interesting but forgettably innocuous. 'Oh, and there was also this in your trap,' Dave says, casually handing over a second pot. It houses the night's first Dusky Hook-tip.

I leap up and bear-hug my friend.

Dusky Hook-tip proves exquisite. Its form follows the same template as Pebble Hook-tips of summer gardens or Scarce Hook-tips of June's Wye Valley night. Curved wings boomerang backwards, parting to reveal the butt cheeks formed by the hindwings. The forewings are strongly cinnamon, generously dusted with iron filings that sheen lilac in our torchlight. Black pencil ripples across each wing – an H grade near the head, boldening to a B at the rear.

The night produces other pleasant surprises – notably the scarce Olive Crescent and another recent colonist, Langmaid's Yellow Underwing – but two further Dusky Hook-tips demand pole position. As 3 am approaches, our exhausted bodies demand that we call it a night.

Will and I return to Orlestone Forest for the following two nights. Again we make for Longrope and are joined by local moth-ers. This time we have twin ambitions: we want to catch Triangle, a target that eluded us in July, and we aim to add Longrope to the expanding list of Orlestone sites that harbour Dark Crimson Underwings this summer – one local moth-er has caught nineteen, another sixteen.

There are more than 200 species of the genus *Catocala* worldwide. All received the same broad memo on dress code: camouflaged forewings in shades of brown, grey, beige and black, with a brightly coloured splash on the hindwings. On the latter some individuality is permissible, with shades of scarlet, orange, yellow and blue; likewise, variation is allowed in the extent of black countershading.

The moths flash their brightly coloured underwings to startle approaching predators: the genus name *Catocala* derives from Greek for 'beautiful behind'. This in-joke made early (and apparently exclusively male) entomologists think of titillating female underwear. Subsequent naming of European species was characterised by a sexist series of etymological jokes. In the mid-eighteenth century, the otherwise great Swedish taxonomist and naturalist Carl Linnaeus initiated the fad, naming two new species as *Catocala nupta* (meaning 'bride'; Red Underwing, the most familiar British species) and *C. pacta* (meaning 'betrothed'). In years to come, the taxonomic list expanded with another bride (*sponsa*, Dark Crimson Underwing); *promissa*, meaning 'pledged in marriage' (Light Crimson Underwing, in British terms a rare breeder and migrant); *elocata*, which equates to 'prostitute' (French Red Underwing, known in Britain from a single record); *electa*, meaning 'fiancée' (Rosy Underwing, a vagrant with less than British records and lately a possible Dorset breeder); and *conjuncta*, meaning 'paired' (Minsmere Crimson Underwing, another single-record vagrant).

The *Catocala* have long been perhaps the ultimate prize for sugarers, particularly because they are only reluctantly attracted to light. Later in the year, Will and I will see for ourselves the physical legacy of Victorian sugaring in the New Forest. At a famed site for both Dark and Light Crimson Underwings, we find old, wrinkled oaks graffitied with long black stripes the width of a hand. These indelible pseudo-burns trace the route of hundreds of treacle-sticky paintbrushes across multiple decades until entomologists were informed in 1905 that the practice was banned in the New Forest. How many Crimson Underwings were lured to sup here before being sequestered into private collections?

Virginia Woolf writes of sugaring in her essay 'Reading', recounting a night-time walk that culminates in excitedly approaching trees whose trunks she and friends had earlier pasted with sweet-smelling liquid. The trick worked;

moths were feasting avidly on the rich sugar. Woolf remarks upon the depths to which each moth's proboscis was plunged and perceives ecstasy in their quivering wings. Woolf's thrill peaks when she observes one moth's lamp-like eyes and burning red underwings. It was one of the underwings, presumably Red Underwing but conceivably the rarer Dark Crimson. Not without chagrin, Woolf and fellow moth-hunters moved towards the moth, intending to catch the splendid creature for their collection. But the moth flew, its departure causing Woolf as much dismay as if she had lost a priceless artefact.

For the second time in as many nights, Will and I can empathise. As I gaze with soft fascination into our trap, Will notices an unidentified underwing drifting down from the oak canopy towards our lights and wine ropes. As we rise from our camping chairs, net in hand, the moth bats an effortless upwards return, dissolving into the thickness of night. Gone.

Woolf's group later caught their quarry. As they consigned it to a killing pot to bolster someone's specimen collection, Woolf's pleasure was slashed by a soulful, portentous noise that transpired to be a falling tree. Her message is clear: the group's taking of life has disturbed the ecosystem, driving adverse consequences. Woolf was evidently troubled by the realisation; later that year, she penned an essay entitled 'The Death of the Moth'. Like Woolf, our perseverance engenders success. Unlike Woolf, our collection involves only photos, so the moths live on. And so no tree falls.

We catch two Dark Crimson Underwings tonight. Both ignore the bright MV lights flaring the main path, instead favouring a weaker actinic light hidden deep within the wood's obscurity, placed judiciously below a treacly rope. Around midnight, as I check ropes and traps, scarlet sears my vision. It seems to have headed towards the discarded grey rucksack in which I transport the little actinic trap. Surely not? Gingerly, I ease open the bag. Resting there, vast but tranquil, is our first rare *Catocala* bride. Its forewing is richly toned, chestnut and sand offsetting the lead and mud. The patterning musters a complex jigsaw, with spots and swirls and shadows breaking the moth's form and guaranteeing crypsis. I tease apart its camouflage to reveal the scarlet secret beneath.

An hour later, I watch a near-replay of the first encounter. Only, this time, the moth crashes into the actinic trap. Ferreting inside, I find it skulking under the bottom egg-tray. We can proffer a brace of Dark Crimson Underwings as further evidence that this rarity is recolonising a county it had long vacated.

Half an hour later, on another trap round, we get our desired Triangle. Except we don't. In the base of a trap purposefully left on the root-toes of a mighty oak, Will spots a female of this scarce species. It is tubbily tiny, not quite as geometrical as its name suggests, and gingery brown. We watch her kick a leg, then freeze. And not move again. We realise that we have witnessed the moth's end of days. Although there is no suggestion that our trap has led to the Triangle's death – she is worn and August marks the limit of the species's flight season – our exasperation is inundated with sadness.

A bigger shock fortunately proves more pleasant. At Saltwood we had been surprised to trap a handful of *Acrobasis tumidana*, a distinctive little micromoth with a

Above: The moth that started it all – Poplar Hawk-moth – being admired by my daughter Maya.

Left: January starts, and so does the year's first garden moth trap. The following morning, however, it was empty.

Below left: Mark Parsons of Butterfly Conservation searching scree slopes on Dorset's Isle of Portland for *Eudarcia richardsoni*, an exceedingly rare micromoth.

Below right: Buff-tip, a moth that contrives to mimic a snapped twig at both ends. How is that even real?

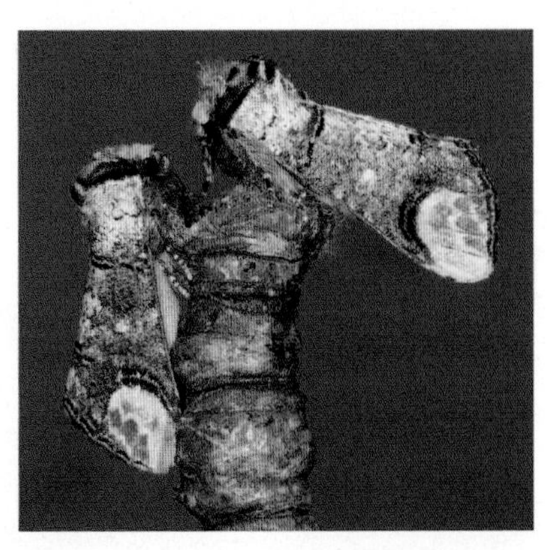

Above: One of the 25 Kentish Glory we encounter over a fabulous Easter weekend in Strathspey and the Muir o Dinnet.

Left: Candy-pink Elephant Hawk-moths, an exquisite and common garden moth, with their No 1 admirer Maya.

Left: The bonsai jewel formally know as *Alabonia geoffrella*, but which many moth-ers know as 'Geoff'.

Below: Clearwing moths excel at mimicking wasps. A giant among thei British contingent is the Hornet Moth

Above left: Britain's most remarkable moth rediscovery of 2019 was a population of the stunning *Hypercallia citrinalis*, which transpired to be hidden in plain sight on Kent's chalk downland.

Above right: Moths don't need much in the way of living space. Led by Sharon Hearle, Butterfly Conservation volunteers search a narrow, Flixweed-smattered public bridleway between Suffolk cereal fields for the rare Grey Carpet.

Left: A Puss Moth luxuriates on the finger of Marna, a University of East Anglia undergraduate participating in the University Moth Challenge.

Below: Our greatest personal discovery during 2019 was this female Marsh Moth in Lincolnshire – the first of her sex seen in Britain since the 1940s.

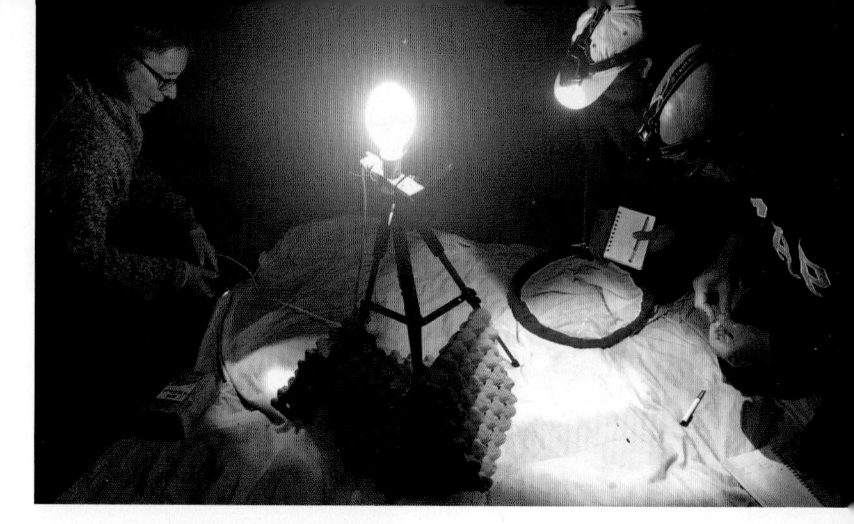

Above: Moth-ers at Orlestone Forest, examining multitudinous creatures settling on the white sheet and eggboxes below a searingly bright mercury vapour lamp. Left to right: Sian Tempest, 'Wingman' Will Soar, James Hunter.

Left: After 38 hours without sleep, Will and I yomp uphill to track down an isolated colony of Transparent Burnets on an Oban summit. It proves the first of a trio of rare burnets we see during three very fortunate days.

Below left: My daughter Maya keeps her own record of our garden moths.

Below right: Author somehow carrying five moth traps about his person, en route to trapping on the Norfolk coast (Will Soar).

Above left: The ultimate in moth infrastructure: the new 'moffice' at Sandwich Bay Bird Observatory, Kent.

Above right: Some catches beggar belief. Dungeness Bird Observatory warden David Walker estimated 15,000 *Synaphe punctalis*, a micromoth of dry coasts, were crammed into just two traps set alongside the observatory building.

Right: Jen Nightingale and team turn into mountain goats to survey the rare Silky Wave on the steep slopes of Avon Gorge.

Below: Another 'how can this moth be real?' moment. Adult *Cynaeda dentalis* are a dead ringer for a grass head.

Right: The toughest night's mothing of them all. We can only get one car up the steep track at Sychnant Pass – and it wasn't mine. So we have to lug all our kit uphill from the main road.

Left: Moth-trapping can be exhausting, particularly in remote places. Here at Sychnant Pass in Conwy, the detritus of a sleepless night's mothing is littered across our base camp like the aftermath of a really hectic party.

Below left: An old specimen stuck to a card held by Durwyn Liley is as close as we come in our attempt to refind Speckled Footman.

Below right: The author carrying kit onto heathland at Wareham Forest in an attempt to rediscover the possibly extinct Speckled Footman (Peter Moore).

Above left: Checking the trap at sunrise in a heatwave at Dorset's Slepe Heath.

Above right: Sienna and Matthew Alwan, former neighbours turned moth-ers, are among many children enchanted by these underrated insects.

Left: Examining migrant moths caught at Suffolk's Bawdsey Hall on National Moth Night. Left to right: Matthew Deans, Jack Morris, Mark Hows.

Below: Jersey Tiger (left) and Gypsy Moth (right): two new arrivals in London, albeit through different routes.

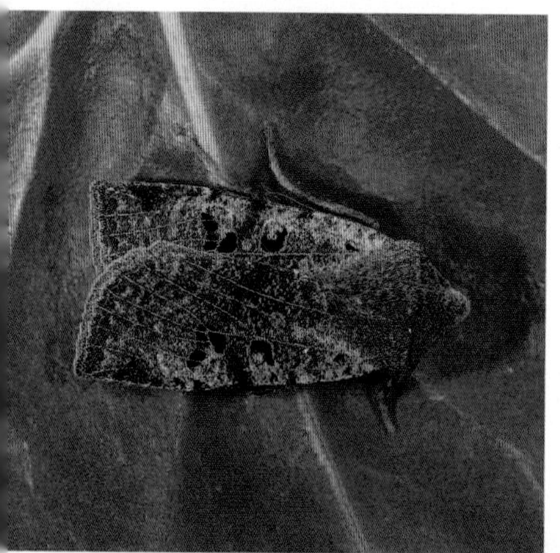

Above left: The consummate migrant moth: Oleander Hawk-moth, caught by Chris Fox in north Dorset.

Above: The brick-sized Clifden Nonpareil, spotted loitering by a light at Brockenhurst railway station.

Left: Black-spotted Chestnut: not a looker, but the moth that my mentor James Hunter helped usher onto the British list and a personally resonant culmination to the year.

Below: A star moth of autumn gardens: the apparently lichen-sprouting Merveille du Jour.

spiny ridge of rusty tufts on the front half of its wings. This big-eyed knot-horn is an annual immigrant to Britain, but no more than forty-eight have been seen across any one year. In our first night at Longrope, we rack up a record-shattering ninety-two. Incredulously we check and double-check our identification. Most intriguingly, we catch no other migrants, raising the possibility that these moths are locally bred, and that – since many look very fresh – we have chanced upon a mass emergence. Might Britain have yet another new colonist, secreted in its forested midst?

Netted Carpet is one of those moths that has it all. Lying flat, its night-coloured wings mesh intricate etchings with pearly scribbles whose swoops and twists draw gasps from those lucky enough to see it. Netted Carpet is passionately rare, considered nationally endangered. To court it, you must invest energy, time and carbon: it survives only in a confined area of Cumbria and northernmost Lancashire. Its recent population surge in the Lake District – a 900 per cent jump in larvae counted since the turn of the millennium – is down almost entirely to a quarter-century of devoted conservation action led by the National Trust's John Hooson. But for us, it is a moth that is just not meant to be.

We join the moth in putting our eggs in a single basket. Netted Carpet is wholly dependent on Touch-me-not Balsam, a yellow-flowered plant as environmentally sensitive as it is delicate. For the moth to survive, its sole larval provender must thrive. And for us to see Netted

Carpet in a woodland near the Lancastrian home of moth-er Brian Hancock, it needs to be flying on the single night that we have available for a 600-mile round trip. Whatever the weather.

Given its cramped British distribution, Netted Carpet unsurprisingly eluded detection until 1856. The moth's rarity and restricted range afforded it high value among Victorian entomological traders – some even rearing specimens to supplement wild-caught supply. Although collecting will have done little to help the moth, Netted Carpets don't make things easy for themselves either. It is existentially risky being dependent on a single scarce and fickle plant.

In 1990, a survey of known Lake District locations revealed a worrying 56 per cent decline in Netted Carpet colonies over ten years. This galvanised research and conservation activity. It was revealed that domestic cattle help Touch-me-not Balsam – and thus Netted Carpet – by trampling competitor vegetation, opening up ground to enable germination, and transporting seeds in muddy hooves to found new colonies. Harnessing of bovine power has worked wonders. Cumbrian woods where Balsam numbers were initially in the hundreds and moths rarely reached double figures were transformed into counts of 124,000 plants and 1,290 caterpillars by 2016.

For 150 years, Netted Carpet was a Cumbrian exclusive. Brian Hancock enters the story upon the moth's unexpected discovery in Lancashire during 2009. We arrange to join Brian in a mature, leafily lime-green woodland in the shadow of the limestone pavements of Warton Crag and Gait Barrows. Here he has dedicated the past decade to monitoring and assisting both plant and moth. Sadly, the rarities have not repaid his devotion. Balsam is being outcompeted by rank, nutrient-guzzling vegetation favoured by ever-milder winters. Where there were nearly 5,000 plants in 2013, Brian laments, barely a hundred remain. Fewer than fifty Netted Carpet larvae were counted in 2018, under a sixth the total of five years previously.

We can afford only a single roll of the dice for Netted Carpet. A few days before we are due to visit, Brian catches his first of the year: the omens are positive. But as the dawn of our departure leadens the sky, we learn that a thirty-six-hour downpour is expected to drench our visit. We set off nevertheless.

An hour down the road, Brian phones. It is bad news. 'This morning's rain is predicted to get heavier and last all tomorrow too. I am sorry but I really think there's no chance of Netted Carpet tonight.'

Will and I sit and fret and swear. We fiddle with phones, checking different weather forecasts and rain radars. To make it worth the effort, cost and time, we need a single dry-ish hour at or after nightfall. But every forecast concurs: unstinting, heavy rain. We admit failure and turn round. Some things are not meant to be.

The bitterest moment is yet to come. Obsessively, I check the rain map throughout the day. Mid-evening, the forecast for Lancashire suddenly changes: shortly before dusk, the rain will abate for an hour, perhaps two.

The realisation pierces me. Had we persisted with our journey, we would have been rewarded with the conditions necessary for our quarry to be airborne. Of all the sylvan moths on our year's agenda, Netted Carpet is perhaps the most clandestine. It eluded discovery in Britain until late in the golden age of mothing. It divulged nothing of itself in Lancashire for a further 150 years. And it refuses to reveal itself to us this year. Some woodlands hug their most precious sylvan secrets tightly to their chest.

***Postscript**: in September 2021, Douglas Boyes – the 'flamboyant soul' who had revealed to us the wonders of Wytham Woods – died suddenly. His tragic passing is an immense loss. This paperback edition is dedicated to his memory.*

11

All the moths look the same

Argyll & Bute, and Highlands
June

Late June. Time to execute an audacious Highlands heist, the year's riskiest trip. The quarry comprises a trio of scarce moths – respectively: rare, very rare and obscenely rare – so finding them is a major undertaking. That's if Will and I can discover where they are. The sole British site for the obscenely rare creature is a closely guarded secret. Our dates are inflexible – and suboptimal: early in the season for one moth and late for two. The bookies reckon none will be on the wing. Even if they are, these moths will only fly on windless, sunny days at odds with habitual conditions in westernmost Scotland. To succeed, *everything* has to go right. The ambition of Ocean's Eleven is a shadow of ours.

Moreover, it's an environmentally irresponsible journey. We'll be bedding down 500 miles from home. For folk keenly aware of our carbon footprint, this is anathema. Carbon offsetting will only partly assuage the guilt sullying my veins. It's also a daft quest – one that makes my wife Sharon send out a search party for my marbles. Because we're looking for three species of burnet moth that appear almost identical – not only to one another, but also to three commoner types that occur close to home.

I was once among the many people who believe all moths to be doppelgängers. That may be true for some groups – field guide pages with serried rows of small, brown and boring noctuids try most people's patience – but generally this attitude smacks of prejudice fettered with ignorance. With our targets in western Scotland, however, it's bang on. They may boast black iridescence ('burnet', a name established by the time of Moses Harris's 1748 book *The Aurelian*, means dark, rather like 'brunette') with blazing-red blotches rather than muddy tones, but Transparent, New Forest and Slender Scotch Burnets are superficially dead ringers. Among burnets, for sure, all the moths look the same.

Our journey to the other end of Britain is clearly not a sane one. But that's the allure of the rare. Humans seem to crave rare things because they allow us to distinguish ourselves from others: I own a Matisse, ergo I am richer, more powerful, worthier than you. Pursuing rarity might also imply success: I have all I need for survival, so can afford for my desires to rest on perceived needs, rather than actual ones. But there are downsides. In his essay 'The Blinded Eye', John Fowles argues that rarity chasing is a mode of destruction, although he considers that what might actually end up being destroyed is less the rare object of the quest than its egotistical, obsessed pursuer. Furthermore, the concept of rarity is sometimes simply in the mind. In one experiment, posh Parisians tasting two caviar samples greatly preferred the one they were told was rare – even though the samples were identical. In another study, people spent more time trying to download an internet slideshow of 'rare' animals than they did of 'common' ones. For conservationists, this levies a risk. The value humans place on rarity – and the consequent overharvesting of wild creatures – could cause the extinction of rare species.

Within Britain, all three targets are assuredly rare. Two are threatened with national extinction – a sorry claim to fame. New Forest Burnet is among only eight British moths classified as critically endangered and is legally protected. Slender Scotch Burnet is just one rung down the ladder of vulnerability; plus, thousands of years of isolation have led to the evolution of locally endemic subspecies. Our particular versions of the three burnets live nowhere else in the world.

Over and beyond their rarity, burnets are special. To the confusion of early entomologists, they blur whatever nominal boundary exists between butterfly and moth. These brightly coloured day-flyers sip nectar and boast butterfly-like clubbed antennae yet rest like moths with tented wings. The shocking crimson colouration warns would-be predators not to munch them. The moths are full of hydrogen cyanide – the consequence of caterpillars metabolising toxins in their foodplant – which males gift to females during mating. Equally remarkably, burnets also take out an insurance policy against the vagaries of Scottish summers. Some of each year's caterpillars overwinter twice, spreading the risk of the whole population being eradicated by inclemency.

Burnet moths may be closely related, look very similar and share an approach to honest communication of their toxicity. But what makes them particularly fascinating are their differences. Each is a distinct evolutionary currency, harnessing a disparate niche in the quest to thrive. Each has a different combination of pernickety demands: the precise incline of slope, the appropriate level of grazing or the right amount of bare ground. Each to its own.

For our burnet saga, we have an eight-hour drive ahead of us and the weather is shocking. Windscreen wipers scamper without respite, roads sheet endlessly with water. But as we draw parallel with Oban, ten miles from our first destination in Argyll, the remarkable happens: the tap turns off, the skies clear and the sun emerges. As we hem Ardmucknish Bay, Beinn Lora swells above us. Deep scars incise grey diagonals across the siskin-green before a spray of scree announces a sheer ascent to the peak. In absolute terms, Beinn Lora is a hill not a mountain. But we'll be ascending from sea level – so its 300 metres are all ours.

We cut the engine at 6 pm. Even in Argyll's first sun for two days, that's late to start climbing and searching for the day-flying Transparent Burnet. Particularly when the information we're using – a series of grid references marking self-contained colonies – is nearly a decade old. My optimism wanes.

We traipse up recently felled switchbacks, embalmed in sweat amid the unexpected warmth. Vehicles below shrink to Lego cars as we discuss prospects between gasps for breath. Transparent Burnet's British range is constrained to steep coastal slopes in Argyll and the Inner Hebrides. Seeing it requires commitment.

Parting company with the trail, we strike out over undulating, heathery grassland. The weather has hardened; it is now breezy and cool. Will reminds me to look for the ground-hugging foodplant, Wild Thyme. We follow a compass bearing to a set of ten-figure grid references where, once upon a time, Transparent Burnets lived. This dumps us at the crest of a south-facing scar that we noticed from the coast road. We have erred and must recalibrate.

We stumble across another grouping of co-ordinates. The bobbly, sloping ground looks just the same as everywhere else – dense, tussocky grass, free of Thyme's rosy-flowered serenade. This too is wrong. Transparent Burnet favours unstable slopes, where regular slippage prevents grass becoming tufty. The longer we search forlornly, the bolder dusk becomes, further corroding our chances. Although largely the colour of night, burnets are not fans of the dark.

We reach a third suite of waypoints and immediately notice differences. Above us a lead-grey cliff towers, its skirt of scree suggestive of mutability. Below us the land slips sharply. But we stand astride a cricket-pitch strip of flat ground, the vegetation buzz-cut short. Flowering herbs – including Tormentil and Bird's-foot Trefoil – abound. 'There's Thyme too, Will,' I cry. Lots of it.

'Here's one!' he replies jubilantly. 'And another!' Will has found the colony of Transparent Burnet. They are sumptuous. A fluffy black head trails backwards like a 1970s mullet and rumples forward onto the underside of the thorax to form a snug onesie. Bloody crimson leaches across three-quarters of the wings, leaving only a broad fringe to be acquired by burnished blackness.

We count fifteen individuals crammed into a rectangle under thirty metres square. Most perch on protruding stems of grass or Thyme, angling damp wings towards where the ebbing sun ought to be, beseeching its reappearance. One pair has summoned the energy to couple; the female's generous abdomen is ringed white. A second male tries to wheedle his way in, but she's not contemplating a threesome.

The moths are fresh and most Thyme remains to bloom. Our visit seems to have coincided with the very start of this tiny colony's adult existence. We could not have timed things better.

After indulging ourselves with a whole seven hours' sleep, we rendezvous with legendary moth-er Dave Grundy on a wild stretch of Argyll's coastline. Bespectacled and bearded, Dave greets us with shy excitement, which is saying something for a man who travels away from home to go mothing on 200 nights a year. We have been granted privileged access to Britain's sole site for New Forest Burnet, a clandestine location where visits are normally prohibited. Tom Prescott of Butterfly Conservation Scotland is happy for us to witness how this organisation and others, including the Burnet Study Group and Scottish government bodies, have worked with the benevolent landowner to save this particular moth from the precipice of national extinction.

As the name suggests, Britain's inaugural New Forest Burnets were discovered in Hampshire in 1869. By 1927, however, degradation of the species' habitat – heathland edge and wooded clearings – and, particularly, exhaustive collecting had caused its extirpation, with professional, care-not collectors removing all adults that emerged in a season. Thanks to obsession-fuelled greed, Britain's New Forest Burnets were no more.

What happened next was unforeseeable. In 1963, an intrepid soul called F.C. Best chanced upon a previously unknown colony of New Forest Burnets on a south-facing grassy sea-cliff more than 500 miles away. He had been looking for Transparent Burnets but got far more than he bargained for. The find proved to be a hitherto undiscovered subspecies, named *argyllensis* to reflect its location.

Tom had warned that late June might be too early to see New Forest Burnet. He feared they might remain encased in papery urine-yellow cocoons clinging to low vegetation like a throaty gob of spittle. The books agree, decreeing the flight season to be July. But we have a heavily constrained window of opportunity: it's 26 June or never. At least we strike it lucky with the weather. The elements are unexpectedly favourable as a wholehearted sun fries us from an unblemished flax-flower sky.

Reaching the New Forest Burnet's domicile demands a mini-expedition. Hiking downhill over rough ground for ninety minutes, we circumnavigate bogs over which Golden-ringed Dragonflies maraud, then follow a stone-littered shoreline that fringes a glistering sea. We pick our way carefully over boulders then wade through chest-high, tick-plagued bracken. Puffing steeply up the faintest impression of a sheep track, we gain the first of two imposing anti-ovine fences. Beyond this first deterrent, any semblance of path dissipates. We contour, step by step, with a jutting cliff to our right and sea-smashed rocks but a short tumble away. Unscathed, we make it to a second barrier and pass through to reach hallowed ground.

Unexpectedly, a dog rounds the corner towards us, followed shortly by its master. The man pulls up short, as do we. Neither of us is expecting to see another person today. At a secret site, where access is by invitation only, we each suspect the other an illicit interloper. Our respective approaches are wary. But all is well. Our mystery man is Neil Ravenscroft, an entomologist contracted by Scottish Natural Heritage to survey New Forest Burnets this year. Our explanation similarly garners Neil's acceptance, and we all relax.

Bumping into Neil means we gather real-time intelligence on whether the moths are out yet or whether Tom Prescott's warning proves prescient. Neil beams his response: 'You're in luck. I've seen five this morning along the top of the slope. Given recent weather, they probably emerged today. Just walk beyond that next cliff. You won't have any problems seeing them.'

Nervously we edge round 'the next cliff'. A spine of smoky rock charges coastwards before being supplanted by sulks of supine basalt. A peregrine soars overhead; a Dark Green Fritillary butterfly blazes its own trail; Meadow Grasshoppers stridulate from our feet. Stretching ahead is a scallop of steeply sloping cliff-edge grassland. Barely mustering a single hectare, it's a tiny area to house the entire national population of a species. And yet a huge realm to search for a moth the length of my fingernail. Needle, haystack.

'Here's one!' It's Will again. 'And another!' Always Will. We've barely stumbled five metres beyond the cliff and he's found a New Forest Burnet.

We collapse in kneeling reverence as the moth uncurls its helter-skelter proboscis then inserts it into the nectar font of a Thyme flower. The insect appears unperturbed by our admiration. It is dinkier than other burnets, with five clearly circumscribed carmine spots, the third small and elongate. There's an ink-pad texture to its blackness, the hint of sheeny blue dusting.

Thirty years ago, such a straightforward encounter would have been unthinkable. In 1989 entomologists clocked that nobody had seen New Forest Burnets for several years, prompting concern that it had again become extinct. To lose a species once might be regarded as misfortune, but to lose it twice would look like carelessness. Fortunately, the moth was rediscovered in 1990 – just in time. Neil got involved soon afterwards and recalls that all fifteen adults he saw one summer were crammed onto the narrowest of cliff ledges, the only place spared the unconstrained nibbling of sheep teeth. Conservationists beavered away, fencing the area to prevent the unwanted incursion of sheep. Unlike a fated attempt in the 2000s to create a satellite colony nearby, the gambit worked: by 2012, the population reached 12,500 adults. Panic over.

Unfortunately, the Argyll coast is a fragile, unforgiving and dynamic place. In November 2014, a landslide smashed the fence. Sheep didn't hesitate, ingressing and browsing to their stomachs' content. In summer 2015, transects suggested that New Forest Burnet numbers had crashed by 90 per cent. The peak count was fourteen moths. Butterfly Conservation Scotland volunteers conducted emergency repairs, only for the following winter to impart further damage.

A proper new fence was required. Given the unforgiving terrain, sourcing a contractor proved tricky. Finally local fencer Seumas MacNeil rose to the challenge. The site's remoteness meant materials had to be helicoptered in. Even then, wind and rain delayed progress. Eventually Seumas erected a new barrier inside the old one, hopefully securing the site. So far, so good. During the 2018 heatwave, numbers soared by 60 per cent, the highest count since 2013. But what of this year?

We stumble slowly westward, trying not to slip on scree confetti. We count nineteen New Forest Burnets emblazoning pockets of Thyme, sun-yellow Meadow Vetchling and golden Bird's-foot Trefoil. There could be more; adults are busying themselves all over. On the return, we find at least fifteen. But there could easily be fifty airborne. Today truly is the first emergence day. Again, our timing seems to be impeccable.

Ambling back to the car, we agree that today has been something else. We feel honoured to visit the New Forest Burnet colony, blessed by the beauty of both moth and its home, and in awe of conservationists' endeavours to reverse its fortunes. We celebrate by adjourning to a local hostelry.

Over an ale or three, Dave quietly recounts his mothing life. Originally a birder, he was intrigued by seeing moth traps run on the Isles of Scilly, then properly enthused by a Walsall Council colleague. This year marks Dave's silver anniversary of mothing. He now runs courses deepening people's appreciation of moths and leads mothing holidays in Spain. But above all, he is impelled to break new ground – operating moth traps in places that nobody else tries, yearning to discover what lives there and to share that knowledge. We meet no more dedicated moth-er all year. Genial, unassuming and modest, Dave imparts wisdom freely. Will and I quickly warm to this understated hero.

A little sore of head the next morning, we traipse out of our shoreline bothy to check traps secreted overnight. A camper observes us examining the catch and bumbles over, intrigued. 'Are those *moths*? Have you caught much?' We show him Poplar Hawk-moth and Peppered Moth, the former astonishing him with its might. The latter inspires recollection of school biology classes; he's never seen this evolution lesson-plan in life. 'Amazing!' he says. 'Moths on a Scottish beach in the sun. Not how I expected to start today.' We bid our new friend farewell. We have a ferry to catch.

To claim the burnet Triple Crown, we cross to the Isle of Mull. Although Slender Scotch Burnet now resides at just five sites there plus one on the islet of Ulva, we are confident. Not only is the weather perfect – a second consecutive day of high summer – but Dave is taking us to where he amassed eighty-odd two days earlier.

Glengorm Castle may be rather grandiosely named for a nineteenth-century country house, but the sharply turreted building commands coastal grassland that slides into the Atlantic. The sun flares silver as we yomp downhill to the base of an upstanding rocky protrusion flecked with Thyme and Bird's-foot Trefoil, where Dave suggests we search.

It was only during surveys for Marsh Fritillary, a butterfly, in 2010 that conservationists realised Slender Scotch Burnet was even at Glengorm. Once we get our eye in – realising that Slender Scotch Burnet is smaller than the familiar Six-spot Burnets – we find our wee quarry *everywhere*. An impressive-enough sixty Six-spots are eclipsed by 537 Slender Scotch on and around the basalt plug. It's hard to imagine anyone ever counting so many. Slender Scotch garb is midway between that of Transparent and New Forest Burnets. There are nominally six red spots on the wing, but the outermost two fuse to form a kidney and the foremost swishes backwards to a point, as if yearning to imitate Transparent.

As with New Forest Burnet, Mull's special moth needs carefully managed grazing to prevent vegetation from becoming rank. Its fortunes are thus interwoven with traditional agricultural practices. Like Transparent Burnet, Slender Scotch demands short turf and bare ground, which may partly explain why it cleaves to rock and soil slides. Such stringent conditions render rarity inevitable. Furthermore, it's a home under threat. Away from Glengorm, invasive cotoneaster plants have blanketed one Slender Scotch dominion, prompting Butterfly Conservation Scotland to deploy herbicide-spraying contractors and secateurs-toting volunteers. On its Burg estate, the National Trust for Scotland has used a helicopter to allay the relentless spread of bracken over the burnet's favoured cliffs.

Back at Glengorm cafe, we learn of another insidious threat. Alarmed by Will's sweep-net, a ranger interrogates his motives. Had we come to steal burnets? Aware that two collectors were once apprehended with hundreds of Slender Scotch Burnets in their van, swiped legally but immorally from Mull, Will explains our honourable intentions – to watch, count and admire. The ranger is satisfied but cautions that clandestine collecting continues to threaten Slender Scotch Burnet. That moth-rustling remains a danger for this species as well as New Forest Burnet chastens us. When human pursuit of the rare deprives a creature of its right to exist, it's time to stop.

We have to stop too, but for a different reason: our time is up. Fortune has favoured the audacious. Our Highland heist could not have gone better. All three species, in numbers. It would be wonderful to wrap up Britain's burnets by joining Dave in trying for Talisker and Mountain Burnets elsewhere in Scotland. But we lack the time and, it turns out, Dave lacks the weather. A reversion to wind and rain scuppers his attempt.

Dave doesn't seem overly bothered by missing out. Nor would I be. I discover that I lack the completer-finisher streak that marks out true obsessives. For me, the allure of the rare is not about marking myself out from others. I'm simply not competitive enough. For me, the excitement of the rare instead lies in its novelty; I don't know what to expect. It also stems from the thrill of the chase, to freighthopping uncertainty and accepting the prospect of failure. But, above all, my pursuit of these and other rarities is about the entire experience. All these moths may look the same, but the settings in which we see them, the conservation efforts they have inspired and the stories they incant are equally defined by their differences.

12

The summer garden – and its lost souls

Norfolk
June–August

At a BBQ in a cottage garden neighbouring Norfolk's secondary estuary, I get chatting to a fellow parent as our children alternately hide or seek. She transpires to be a moth-er as well as a mother, and our conversation encompasses engaging kids with nature and the lepidopteran stand-off. 'Butterflies have it easy!' my companion complains. 'Moths are much better, far more interesting. And they're right there – in our garden. Our kids love moths.'

The conversation surprises and delights in equal measure. It is forty years since the ecologist and author Robert Pyle conceptualised 'the extinction of experience'. Urbanisation, he argued, was reducing personal contact with nature, breeding apathy towards environmental concerns, thereby countenancing further habitat degradation and 'sucking life from the land'. With local extirpations, he wrote in the journal *Horticulture*, 'the power of the neighbourhood to fascinate, arouse, excite and stimulate also passes into dullness, ennui and apathy'. In turn, we become progressively inured to loss. What, asked Pyle, 'is the extinction of the [Californian] condor to a child who has never known a wren?'

As Mark Cocker eloquently posits in *Our Place*, we are indivisible from the natural world, for *Homo sapiens* – even long-term prisoners detained in solitary confinement – is part of an ecosystem. Yet somehow we are becoming blind or resistant to the truth of biophilia, E.O. Wilson's theory of the innate and genetic affinity between humans and nature. The loss of knowledge of species other than our own scares conservationists. We increasingly talk of 'nature-deficit disorder', gesturing towards the growing body of evidence that demonstrates the significance for our physical and mental well-being of green space inhabited by species other than *Homo sapiens*. Without awareness of something's existence, how can we possibly care enough to make choices to conserve it?

Summer is the season of garden-moth plenty. Maya and I run the trap on twenty June nights, leaving light to weave its magic even when I am out in the sticks. Only four nights fail to produce more than a hundred moths. July is better still: never under 150 moths, with more than fifty species two nights in every three. Balmy nights in late August are the headiest of all. With nearly 700 individuals of seventy-odd species, we don't know where to focus.

This richness merits a youthful audience over and beyond my daughter. For sure, Maya and I still race out of bed every morning as if it were Christmas, eager to see what the Moth Santa has brought. Granted, we still compete: who can guess how many hawk-moths await? And, with every morning, my notebook crams further with Maya's expansive handwriting as she transcribes the creatures we jointly name. But we both crave kindred spirits with whom to share the bounty.

First up are Kitty and Matty, five-year-old twins of friends staying one weekend. They delight at a White Ermine. 'A fluffy snowball,' Kitty suggests. She worries when the moth lies on its back, tiny legs in the air – until I show them that it is pranking us by playing possum. Kitty's parents sigh with relief, the prospect of overwrought child avoided.

Maya's ninth birthday party provides a captive audience too, but it also evidences the disconnect from nature that grows with age and burgeoning screen-time. When Maya turned six, her partygoers spent an hour enthralled by a moth show-and-tell. Eyes wide open, jaws dropped. Three years later, Maya's reassurances that the moths neither sting nor bite fall partly on fallow ground. The cohort has split three ways: into the captivated, the disinterested and the squeamish-fearful. We haemorrhage two acolytes when the Poplar Hawk-moth Maya has posed on her nose excretes pond-green yuckiness over her chin.

But all is not entirely lost. Maya's bestie, Felicity, comes for a sleepover, leaping at the chance to open the trap before Sunday breakfast. She is already a convert, smitten by moths drawn to her porchlight. Today Flic spots her favourite moth sprawling over the egg-trays. That she *has* a favourite moth gladdens me deeply. Black Arches is large and spade-shaped, its white wings are dotted like a leopard, banded with an Andean mountain range and flustered with arrowheads. Other moths are less easy for Flic to identify, so I gift her a copy of a moth field guide written by Sean Clancy. Handing over the book, I feel like I am persuading her to join an Alpha course. Moths are my church.

This matters because moths matter. For a start, we need moths. Out of the sight of day-active people – and thus out of minds too – moths pollinate wildflowers and crops, thereby playing integral roles in a healthy, productive countryside. In one study of Norfolk's agricultural heartland, nearly half the moths tested were transporting pollen – and from a wider range of plants than headline-grabbing bees. Moths also nourish birds, bats, Hedgehogs and more: UK Blue Tits eat an estimated fifty billion moth caterpillars each year.

Remove moths from the food chain and moth-munchers suffer. Britain's Cuckoos are not only declining but moving north and uphill, apparently in response to the agricultural change that has caused plummeting populations of moths such as Garden Tiger. The decline of bat populations on farmland is correlated with dwindling moth numbers. The sluicing of the countryside with chemicals is increasingly thought to be driving moth declines over and beyond the removal of their habitat. In a dangerous irony, farmers are killing off the moths they need to pollinate their crops. One experiment showed that treating plants with nitrogen fertilisers increased the mortality of common moths such as Straw Dot and Blood-vein – both frequent visitors to our garden, where we have not deployed inorganics since assuming ownership.

It follows that moths are indicators of environmental health. Species with narrow ecological niches – a single larval foodplant, say, or a specific microhabitat – offer us a granular early warning system. Alternatively, the alteration in status of an ecological generalist – a species with a wide choice of foodplants and domain – can speak volumes about broad-brush ecological change.

This matters all the more because moths may even have value that we have not yet computed. Astonishingly, two moths that Maya and I catch in our garden this summer – Wax Moth and Indian Meal Moth – have recently been discovered to be capable of feeding on and breaking down polyethylene and polypropylene. Might moths provide the germ of a solution to the global plastics problem?

Even at garden level then, Maya and I are privy to moths' worth – and to their flux. We discuss which species are enjoying a good year. Heart and Dart – nicknamed 'monobrow' because of a black band above the eyes – is thriving, forming two-thirds of some catches. Treble Lines and Dark Arches are notably abundant. But where are all the Small Square-spots? Moth-Twitter is full of nationwide concern at single-figure catches or – worryingly – complete blanks. However, leaping to conclusions based on a few weeks' evidence would be hasty. Populations fluctuate naturally in tandem with weather conditions, relationships with predators or parasites, and so on. Conclusions should only be drawn on longer-term datasets, notably those assembled by Butterfly Conservation's county moth recorders or generated by the Rothamsted Insect Survey, which has compiled half a century's worth of nightly counts from sites nationwide.

Number-crunching by clever folk reveals a wealth of storylines playing out in our garden. Often splattering our house walls is Least Carpet. During our first Norfolk summer, we caught a handful of this pretty, glossy-winged lace curtain. This year we see scores. This fits with an England-wide trend: numbers have increased a hundred-fold since 1970. Least Carpet is a winner.

So too, unexpectedly, is Small Ranunculus. This silver, charcoal and yellow moth resembles a chunk of lichen-encrusted wall – a disguise we test when catching one in August. In 1884, Edward Newman described it as 'abundant in certain localities'. Up to 1900, this noctuid moth was widespread in southern England, even being considered a pest of lettuce, its foodplant. Then, abruptly and mysteriously, it vanished. There were only four records countrywide in the forty years after the Second World War. Bewildered by its apparent extinction, E.B. Ford argued that neither over-collecting nor human meddling with the countryside were responsible. In 1997, entomologist Jim Porter observed that the moth was also declining in northern Europe, suggesting climatic variations and land-use change to be contributory factors.

Miraculously, Small Ranunculus has recolonised Britain. Two friends had insights into its resurgence, as they caught one a year after its rediscovery. In 1998 Ann Duff had just started moth-trapping under the guidance of her now husband, Andrew, in Kent. Working through the field guide, she identified one mystery moth as Small Ranunculus. An experienced moth-er who knew full well that the species was no more in Britain, Andrew pooh-poohed Ann's identification. 'Then I looked at the moth, then at the book, then back at the moth – and realised that Ann was right,' Andrew remembers. 'I had been totally condescending – and wrong!' Small Ranunculus has since spread rapidly across England, reaching Norwich by 2009. The reasons for its return are as unclear as those driving its evanescence. Had it been closeted, hermit-like, in Kent all along – or might imported lettuce seedlings have fermented colonisation? Either way, Small Ranunculus is back.

Many species are not so fortunate. Maya and I giggle at a Mouse Moth – a nondescript, easily overlooked critter – that scurries for seclusion like its rodent namesake rather than flying. But our humour fades when we learn that its abundance has decreased severely and its distribution shrunk significantly. By late 2019, it will have been catapulted onto the Red List of nationally threatened moths.

Even more worrying is Golden Plusia, a nugget of a garden moth boasting a Hercule Poirot moustache formed by upturned mouthparts, a Jane Eyre bonnet straining above its head, and wings of pure bullion. Britain's first arrived in 1890. In 1967, E.B. Ford described it as Britain's most successful recent colonist, with populations established into northern Scotland nourished by garden delphiniums. Since 1970, however, the Plusia's abundance has halved and its range contracted massively. It is now considered endangered at a national level – one of Britain's thirty-odd most threatened larger moths. Our admiration of our only example this year is infused with sadness.

We regain some cheer by venerating other beauty. Swallow-tailed Moth is exquisite: large and lemony, with vampire fangs protruding from the trailing edge of its wings. We revere the oyster coruscation of Mother of Pearl, admire the vibrant leafiness of Scarce Silver-lines – a moth that, unfathomably, sings at dusk from the oak canopy – and spend moody nights in White Satin.

The latter – a pair of sheeny knickers from which peeps a hirsute thorax – kickstarts our giggling at moth monikers. These prove to be a glorious mash-up of history and humour, poetry and pizzazz – the gamut celebrated in Peter Marren's delicious book *Emperors, Admirals and Chimney Sweepers*. We easily discern the witch-like face that conjures the name Mother Shipton, whose Knaresborough cave I visited as a kid. We catch moths named after other creatures, such as the black-spotted Leopard Moth, the searingly scarlet Ruby Tiger and the charcoal-smoked Alder Kitten. We stitch one catch into a stanza, swirling around our witches' cauldron of a moth trap while intoning: 'Clouded Silver… Coxcomb Prominent… Clouded-bordered Brindle… Rustic Shoulder-knot…'

We hoot at moth-er in-jokes premised on the trickiness of identifications: we catch plenty of Uncertain and, away from the garden, I see Suspected and Confused too. Our imagination is ignited by Flame, Flame Shoulder and Scorched Carpet. Names even inspire the non-mothing majority of Twitterati, with proponents mock-proposing that they are actually all Cocteau Twins releases or Farrow & Ball paint colours or characters from the next generation of Marvel movies. As these wags evince, you don't even need to see moths for the creatures to induce wonder.

One particular June catch prompts me to deepen my acquaintance by visiting another friend's garden. Male and female Ghost Moth vary so markedly in appearance that they could be different species. The female is larger than her suitor, with long wings of clotted cream swirled with strawberry jam. The silvery-white male boasts a salted-caramel mop of hair. 'He looks like Donald Trump,' Maya giggles.

Jovial and bespectacled, my oldest friend Chris Sharpe looks nothing like Trump. When I arrived at secondary school, Chris was an erudite, kindly sixth-former who took me birdwatching. Our paths have crossed repeatedly ever since, and just after sunset today we convene on Chapel Green meadow, which flanks Chris's garden in the mid-Norfolk village of Rockland St Peter. Smaller than a tennis court, this fecund public space – a Millennium Meadow, so barely of voting age – riots with wildflowers. Chris and friends have lovingly stewarded the plot for years, with the support of most villagers. Most, but not all. One household is riled by the community meadow's wildness, perceiving nature's vitality as an unkempt eyesore and demanding its conversion to tidy sterility. 'But it's just nature!' Maya retorts when I relate this attitude the following morning. 'And we're part of nature.' Fortunately, villagers almost unanimously voted in favour of conserving the area, so Rocklands Parish Council, Chris and co. – wildlife heroes to the end – are standing firm.

Chris wants to showcase Chapel Green's disco-dancing Ghost Moths. In almost all British moths, the sexes meet by males tracking down females that are exuding sex-scent pheromones. Ghost Moths do things differently. The boys gather in a lek, displaying aerially while releasing a 'come-hither' perfume that they waft into the air with scent-brushes encumbering their hind legs. This I *have* to experience.

The opal sky coagulates to reach the precise light intensity at which Ghost Moths feel comfortable emerging – a compromise between visibility and predator avoidance. The males' upperwing scales exhibit a meshwork designed, like cat's-eye road markers, to reflect ambient light in near-darkness – essential for catching the girls' attention. But whiteness is visible to birds too, so Ghost Moths try to emerge after avian predators have gone to roost. The window of opportunity is tight, for Ghosts also try to finish body-popping before bats get going. This primitive species needs to avoid detection because it has not entered the evolutionary arms race that grants many moths the ultrasonic hearing to discern a rapidly approaching mammalian marauder.

Bang on cue, at 9.50 pm, the first lonesome male Ghost floats into view. Within three minutes, eight are angel-winging around the arena – an opening flecked by Ox-eye Daisies within a sward of thigh-high grasses. Each spectral moth seems tethered by a thread from which it pendulates, swinging nervously and metronomically to and fro. Occasionally, abdomen curving upwards and forewings beating independently of hindwings, one male barges into another, booting it off the dancefloor.

Our swarm is small – 1,600 have been estimated at one Swedish lek – but perfect. To avoid bats, the males stay close to the vegetation. In Shetland and the Faroes there are no bats to bother them. Here, the trouble comes from gulls and Arctic Terns. 'In the white night of the simmer dim,' author and Shetland crofter Jon Dunn regularly sees the master-peregrinator catching the libidinal moths unawares. In a unique evolutionary response, these insular Ghosts have adopted an array of colour forms, ranging from the ubiquitous white phantoms to dusky shadows. By the end of summer, only the latter have not been picked off by visual hunters.

Unseen by us, two females have snuck into Chapel Green, selected a mate and copped off. We track down the pairs amorously attached at the summit of a grass stem, the male dangling below the female, clinging on to create dear life. The unrequited males continue to 'pendeculate', a neologism coined by Bernard Kettlewell (of Peppered Moth fame). Their flickering phosphorescence will-o'-the-wisps for a further five minutes before these lost souls evaporate just as instantaneously as they appeared. The Ghosts of the Green have bid us farewell.

At some point during this summer of moth, I too become a phantom to my family. I sit alongside the garden traps long after the family has gone to bed, watching the whirling and meandering and crunching and spiralling. After too few hours, I rise with the sun, busying myself with names and numbers before the morning warmth vaporises the catch. Too often I forget what I'm meant to be writing, amnesia being the bastard offspring of sleep deprivation. Afterwards, I steal back under the sheets until breakfast, an unfaithful lover concealing his tracks.

The sensation of adultery is close to the bone. One moth addict I meet confesses that 'the hobby takes over your life, no two ways about it,' but flusters that his partner prefers him to be 'infatuated with mothing than taking drugs or eyeing up other women.' Norfolk naturalist and broadcaster Nick Acheson jestingly trolls me on Twitter: 'Damn you, Lowlife Lowen and your luscious Lepidoptera… I've been clean for so long… I haven't so much as perved a passing pug in decades. I went to Mothaholics Anonymous in my twenties and I've been clean for twenty years, but there's a Yellow Shell in my bathroom and it's so pretty and I'm weak…'

I hear you, Nick, I do. But…

13

Life's a beach

Norfolk, Kent and Suffolk
June–August

A loving family afternoon is punctured by an urgent phone call: Keith Kerr has just stumbled across three Scarce Pugs on the Norfolk coast and wonders whether I might want directions.

Too right I do. There are fifty-two species of pug in Britain – one for each week of the year. Perhaps forty are easily ignored, being identikit and nondescript – greyish or brownish creatures with almost imperceptibly different striping or spotting that nominally tells them apart. One friend suggests sneezing at them so they fly away before needing to be identified. But Scarce Pug is something else. Attractive, distinctive and damn rare, this is one of a handful of must-see pugs. Even better, the only place to realistically see it is my home county. Yet few people have. So I make haste to the beach.

In the late nineteenth century, tax inspector Charles Barrett discovered Britain's first Scarce Pug near Kings Lynn in west Norfolk. A renowned entomologist, Barrett authored a seminal eleven-volume work on British moths and butterflies, *The Lepidoptera of the British Islands*, though he died before its completion. In Britain, Scarce Pug was subsequently found in Essex, Yorkshire and Lincolnshire, but clings on only in the latter – and, even then, at just one site. Otherwise, this moth is all about north Norfolk and its beach-fronting saltmarshes – plus a lonesome foodplant and a panoply of threats.

As I stride towards the sand dunes late afternoon, coastal saltmarsh labours west as far as my eye can discern. Its grey-greenness comprises one of the world's rarest habitats, occupying under one-ten-thousandth of its land surface. The bleakness is deceptive. In Britain alone, 150 types of invertebrate – including nineteen moth species – reside only on saltmarsh, having adapted to its salinity and tidal inundations. Many are nationally threatened, for saltmarsh is being squeezed between rising sea levels and coastal defences. One specialist moth of this habitat is already no more: Britain's final Essex Emerald was seen in 1991, its sole pabulum having been buried during the construction of a flood embankment. With nothing to munch, the moth was a goner.

Scarce Pug risks following suit. Like the extinct Emerald, its caterpillars eat only Sea Wormwood, whose Norfolk locations were mapped by conservationist Bex Cartwright as part of her research into a moth that she styles 'a lovely, fascinating species – beautiful in both its adult and larval form'. Fortuitously, Keith's suggested spot – at the intersection of saltmarsh and dune – coincides with one of Bex's recommendations. The testy ground is smattered with pistachio-toned, knee-high Sea Wormwood, its feathery fronds foresting upwards and bushing sideways. I follow Bex's advice to focus on higher, drier hummocks, where caterpillars are less frequently inundated by the tide. After an hour's forlorn scrutiny, tummying along the mud to place my eyes at moth level, I eventually comprehend Bex's enchantment with Scarce Pug.

Twelve inches ahead of me, perched low in clumpy grass, is a boldly banded gem. Silver, black and terracotta diagonals stretch across this pug's long, rounded wings, which it holds far from its exposed, longitudinally striped abdomen. It is a humbug of a moth. For a pug, it is large. For a pug, it is beautiful. For a pug, it is unprecedented fist-pump territory. Unfortunately, I have nobody with whom to share the moment.

For twenty sun-baked minutes, I follow the moth as it wings surprisingly forcefully between exposed perches, before alighting to angle itself towards the solar rays. Each second is a treat. Scarce Pug is among Britain's 'top forty' most endangered moths, surviving on a single foodplant in a threatened habitat and inhabiting very few sites. 'Scarce Pug,' Bex suggests, 'requires more people keeping an eye out for it, surveying and recording it, and monitoring and protecting its sensitive habitat.' Scarce Pug is a moth in need.

Two Sundays later, my alarm clock hollers at 2.30 am. It is my fifth pre-dawn start in a week, the fifth time I have been banished to the spare room to avoid waking the family. Today I am rising to see some of Norfolk's more widespread beach-dwelling moths.

Large numbers of moths are adapted to a littoral life. They eke out a linear existence in barren, salty habitats, adopting strategies to survive the combined oppression of wind and sea. Most moths' larvae may subsist on more plants than does Scarce Pug, but the vegetation will be confined to sandy shores. Inhabiting such a precise environment means these moths rarely venture beyond a narrow coastal strip. One such target species today, Sand Dart, so writes Mark Young in *A Natural History of Moths*, is 'hardly ever caught in moth traps that are even tens of metres behind the seashore'. These moths will never come to me. To see them I must travel.

Well before the sky has even contemplated first light, I wend across a mist-drenched agricultural plain towards the north-east Norfolk enclave of Eccles-next-the-Sea. Ahead, a lighthouse compels my attention and draws me, a moth-er to its flame. Gradually I realise the beacon is my destination. It is no lighthouse, but Neil Bowman's dune-top alpha moth trap.

Neil is a fellow ex-Londoner and photographer who became interested in moths when a car crash confined him to crutches, cramping his freedom to go birdwatching. 'Before the accident, I kept spotting these wretched day-flying moths while doing butterfly surveys,' he jests. 'So, while limping around, I started cold-searching tree trunks. It was finding London's first Clifden Nonpareil [a giant moth with striking blue underwings] that hooked me.' Forsaking London for the Norfolk Broads, he started moth-trapping in earnest. Then, in 1996, he purchased a house and arable plot behind the dunes near Eccles. In the years since, he has allowed nature to reclaim his four acres.

The changes have been profound. In autumn, waxcap mushrooms – indicators of high-quality, chemically unimproved grassland – flourish here. In summer, birds such as Whitethroats sing from the scrub Neil has planted. Bird and bat researchers intensively study this garden reserve, while Neil's network of moth traps – illuminated nightly by the flick of a single switch – has revealed the presence of more species of macromoth than any other Norfolk garden. A goodly number of his garden moths have been additions to the county list. But his approach is more wholesome than mere listing: across twenty-three seasons, he has compiled an enviable dataset of migrants and notable residents, charting their ups and – increasingly – their downs on a card index before uploading them to the Norfolk Moths website.

This requires dedication. Day in, day out, Neil wakes in the dark to check the traps. 'It's a balance between having enough light to see and the Blackbirds getting to the moths before I do,' he explains. We head out, following cables between bramble thickets and across the dunes. Following an unseasonably cold night, Neil is not optimistic. 'I suspect the traps will be virtually empty,' he grumbles. Neil's concept of 'virtually empty' differs somewhat from mine. He is disappointed by a catch that enamours me, for it includes one apiece of three specialities – beach residents – that have prompted my visit.

The names of two noctuids, Sand Dart and Shore Wainscot, speak of our coastal location. The latter is the colour of dunes at first light, the fulsome tawniness of its wing seared by a single white lightning bolt. Sand Dart is more variable, some paler, others looking dirtier, but all lancing the stunted, dark weapon-mark that gives darts their name. Its caterpillar is a clever thing. Feeding in autumn, it burrows deep into the sand to see out winter. Aware that this season's storms will have changed the sand's level and layout, spring sees the larva briefly worm upwards to the surface. It then retraces its wriggling for fifteen vertical centimetres to reach the perfect pupation depth – neither too exposed nor too far underground for the adult to leverage its way to the open air.

The distributions of Sand Dart and Shore Wainscot overlap along much of strandline Britain. See one and you'll probably encounter the other too. But this neighbourliness also means the duo are disappearing from the same places. The range of Shore Wainscot has shrunk by nearly two-thirds this century; that of Sand Dart by almost 90 per cent. The fate of Neil's third early summer speciality, Rosy Wave, lies between Shore and Sand. This delicate moth's distribution has waned by three-quarters this century. A geometer, its ghostly pale wings are held horizontally like a butterfly, sprinkled with moon dust and devoid of all but the faintest pink tinge that whispers its name. Today it is a phantom in more than appearance, vanishing from Neil's trap before I can get a proper look.

The briefness of the encounter, coupled with a soaring drive to catch my own moths rather than see other people's, is frustrating. A fortnight on, I seek redress. Enticed by a warm evening, Wingman Will, Ben Lewis and I join forces north of Great Yarmouth Pleasure Beach. From the coastal boulevard, we traipse east across dunes bound with vegetation. By the time we reach open sand with scattered clumps of Marram Grass and nodding Sea Lavender, our generator-laden pace has slowed to a weary trudge. But we set up efficiently – just as well, for dusk rushes in unexpectedly.

Within half an hour of nightfall, the strand's seemingly blank canvas reveals hidden portraits. The Marram's feather-duster tufts, so spartan to human eyes, evidently secrete sugar, for they are being lapped by thronged Sand Darts and Shore Wainscots. Rosy Waves are flittering from our feet, phantoms still – but this time acquiescing to perusal. Another sand specialist, the furiously emblazoned Archer's Dart, shares a square inch with a White-lined Dart. A Bordered Sallow surprises; even more so the Breckland-loving Marbled Clover, which shouldn't be here at all. A bevy of scarce micromoths with multisyllabic scientific names incant their presence: *Chionodes distinctella*, *Eulamprotes wilkella*, *Thiodia citrana*, *Anerastia lotella*... Unfurling into the early hours, the littoral litany seems unending, its life unexpectedly bounteous yet irrefutably fragile.

Busy old fool, unruly Sun
Why dost thou thus
Through windows, and through curtains, call on us?

The first lines of John Donne's poem 'The Sun Rising' serve as an inscription on the south-facing wall of Prospect Cottage, film director Derek Jarman's former domicile at Dungeness. Donne's intonation strikes a chord, for it is here in southeasternmost Kent, in the shadow of a stringently protected nuclear power plant, where the North Sea pivots into the English Channel, that Wingman Will and I see more of the year's sunrises than anywhere else bar our own houses. Banished to the end of an anomalous terrace, Dungeness Bird Observatory becomes our home from home, warden David Walker our recurrent landlord.

Imposing in stature and deadpan in humour, David is celebrating his thirtieth anniversary at Dungeness. 'I came here just for a year,' David intones. 'But stayed.'

The Observatory was established in 1952 to study bird migration. Although all things feathered remain its mainstay, David keenly records pretty much anything that moves or grows here. In 2015, he found Britain's first colony of Tree Crickets; during 2019 he and friend Gill Hollamby discovered a shieldbug new to the UK. Mothing has a long history at Dungeness, 'with techniques changing over the years from "sugaring" to Tilley Lamps and now MV lights,' David says. Along with long-term residents such as Barry Banson and Sean Clancy, David has surged Dungeness onto the podium of British moth locations – renowned for both migrants and an exceptional suite of rare residents. By the end of 2018, an astonishing 1,343 species had been recorded here – more than half the British list.

But the shingle fulcrum of Dungeness is a strange place. One that polarises opinion. My wife hates it. 'Too barren, too forbidding, too nuclear,' Sharon says, declining yet another invitation to join me down south-east. In contrast, I love it. The spartan landscape deceives, its ostensible vacuity rampantly secreting special creatures. The endless skies unshackle my liberty. I just keep my back to the power station.

Such disparate perspectives reflect the curious ambivalence that characterises many shingle expanses. These pebbly beaches are shape-shifters, perpetually dynamic – battered by storm and salt, but rewarding hardy pioneer plants that persevere once they strike soil. Shingle resists our ingress: one trudge forward is followed by two steps back. We return the disdain, compromising its naturalness with concrete sea defences, military installations and controversial energy infrastructure. Yet, despite our efforts to control and malign, shingle retains an overwhelming wildness – a power of the real.

EDF Energy's off-limits property aside, Dungeness lacks boundaries, fences or other divisions. Even the famed garden of Prospect Cottage, where magenta Foxgloves spar with violet-blue Viper's Bugloss, dissolves into its surroundings. The sense of communality, of a right to roam, is mirrored by its community spirit. Each morning Dungeness Bird Observatory witnesses an arrival of local moth-ers. Some bear gifts: winged wonders that they trapped overnight and have temporarily encased in a plastic pot. But everyone comes to check what David and his assistant warden Jac Turner-Moss have caught. Offering to boil the kettle for a cuppa is but an excuse.

Over the dozen nights that Will and I spend here between June and September, we become part of the Observatory furniture. Our circadian rhythms mark us out, however. We use the Observatory as our base between night-time mothing sessions on the Dungeness shingle, in the nearby RSPB reserve or at the slightly more distant Orlestone Forest. The morning cuppa we share with its faithful, while David and Jac log the contents of their moth traps, marks our final act before snoring the day away. At Dungeness more than anywhere else, we become creatures of the night – for good reason: there is no more exciting assemblage of scarce moths anywhere in Britain.

Taking pole position on our June visit is the simply named White Spot. This is a chocolate drop of a moth – a chunk of Toblerone, flecked with almond nougat. To the backdrop of cackling Marsh Frogs and insomniac Sedge Warblers, we catch fifteen, then appreciate more in David's traps. Come morning, I demonstrate the White

Spot's consummate disruptive camouflage by playing hide-and-seek with one. I place it on a stony path, then invite Will to discern it. He fails. If Will were a predator, this White Spot would live to see another night.

White Spot is a distributional oddity. It ranges from Scandinavia to Iberia yet, in Britain, is peculiarly confined. Here it breeds at perhaps ten shingle pockets between Devon and Kent. The sole foodplant of its caterpillars are the seeds and seed capsules of Nottingham Catchfly. This slender plant grows abundantly on the grassy hummock that neighbours the Observatory. Here we observe that its nodding white flowers remain resolutely closed by day but open for business at dusk. Emitting a heady, hyacinthine scent, they lure nocturnal pollinators to service the plant's propagatory needs. Although a rarity nationally – formally considered near threatened – White Spot appears common where found. 'It's not unusual to get a hundred per night in our traps,' David says. He reflects, then clarifies: 'OK, I might be exaggerating. But we have caught hundreds in a night.'

Four weeks on, Will and I return to Kent. Again, we head for Dungeness but follow a circuitous route along the county's eastern seaboard. Popping into Sandwich Bay Bird Observatory, warden Steffan Walton allows us to turn fridge-raiders in the spanking-new 'moffice' – an airy, spacious wooden building where moth catches are now processed. We regale in various potted delights, notably Bright Wave, a small geometer that is among eight species headlining the Butterfly Conservation project, Kent's Magnificent Moths.

Counting the concentric, tree-trunk rings rippling across this tan-winged creature whets the appetite for our lunchtime rendezvous with Nigel Jarman. Once we have bimbled the requisite few miles southwards, Nigel escorts us for five minutes from his front door to a rectangle of vegetated shingle imprisoned between main road and beach. The size of a football pitch, the strip is both invertebrate haven and potential ecological trap – a sink-estate for insects.

Right now, it is sprouting and pulsating with life. Flowering plants and grasses push upwards through the stony substrate. From their base issue the mechanical buzzes and scissoring of various grasshoppers and crickets. We track down Grey and Roesel's Bush-crickets, but other species stridulate unseen. Marbled White butterflies float everywhere, each an M.C. Escher print in flickering motion. Smaller embers of butterfly – skippers of some kind – flee before us. This is world enough, even before we clasp eyes on the star moth.

All of a sudden, Bright Waves are everywhere. Every few paces, one zips from low on a grass stem, alights on another a few metres on before shuffling a round to the opposite side. 'They always do that,' Nigel rues. The little flutterers are as shy as anything, fixing their eyes on danger. 'They're impossible to photograph,' Will moans.

Bright Wave breeds along barely ten miles of east Kent's coast – centred on Deal – and at one inland site, but nowhere else in Britain. The moth has a long history around this coastal town. In 1883 entomologist H.J. Harding rambled around Deal one summer's day, where he 'fell in with a small brown moth [Bright Wave] in large numbers'. Although long lost as a breeding species from Essex, Bright Wave populations appear stable at Kent's eight current sites, including Nigel's recently discovered location. 'I've counted up to 177 at a time here,' he says, pleased as punch. His oasis may be one of this subtle moth's most important homes, but its future here is insecure.

'The shingle used to extend seawards,' Nigel says. 'But planning permission was granted before I discovered the moths, and someone built that big house on it.' Sullenly, he gestures towards an adjacent looming property, then points to the ground below our feet. 'The bit we are on now remains owned by the council, but who knows whether there's a risk that they will sell it. We haven't been able to persuade Natural England to extend the nearest Site of Special Scientific Interest to include this place. Too much paperwork apparently.' His frustration is clear. He's been mothing since he was twelve and moved to Kingsdown in 1996. To stumble upon such an important new population of such a rare moth on his doorstep was a dream come true. To watch it be bulldozed would constitute a nightmare.

This stretch of Kent's coast also harbours two recently discovered populations of an even rarer and more immediately beautiful moth, Sussex Emerald. To see it, however, we visit its long-term stronghold, Dungeness. But there's a catch. As with Black-veined Moth, Fiery Clearwing and five other moths, it is illegal to disturb Sussex Emerald. Accordingly, during the moth's flight season, visitors to the Observatory are prohibited from running moth traps on the Dungeness estate. To see

Sussex Emerald, we must rely on one being caught by the Observatory staff or a resident moth-er.

Since first found at Dungeness in the 1950s, the fortunes of Sussex Emerald have ebbed and flowed like the tides. Local ecologist Sean Clancy has led work to monitor populations and advise land managers on conservation measures. 'Sussex Emerald depends on Wild Carrot and does best where there is enough soil to allow a light grassy sward to develop,' he explains. The power station complex aside, however, Dungeness has open access. This is a boon for visitors, but constant footfall impedes soil development. Furthermore, the peninsula's abundant Rabbits are nibbling Wild Carrot before it can grow tall enough for Emerald caterpillars to feast on it. In East Anglia's Breckland, this mammal's absence is causing conservation problems. At Dungeness, its presence is the issue.

In 2008, Sean led the creation of experimental habitat plots. Wild Carrot did best where artificially seeded and fenced off to exclude Rabbits. By 2016, several landowners had collaborated to seed six plots and fence four. That spring, larvae were found on three plots that were both fenced and seeded. Given that counts of caterpillars had suggested a 70 per cent decline across the century's first sixteen years, the essays provide much-needed hope.

One surprising Sussex Emerald stronghold comprises a small back garden marginally north-west of The Pilot Inn, where Will and I routinely consume our sole meal of the mothing day. The garden's owners, Dorothy and Robert Beck, are proud to host the rarity. 'It's an honour to have such a rare species in the garden – and a pleasure that I look forward to each year,' Dorothy explains. Now retired, she is a keen naturalist who has done much to help both Sussex Emerald and a dandelion-like plant, Stinking Hawksbeard, which survived unnoticed at Dungeness despite succumbing to extinction across the rest of the country during the 1980s.

'When that gleaming green beauty reaches our moth trap, I feel both satisfaction at a job well done and relief that all is well with our little part of the world,' she says. For now, Dorothy can relax. But – as with Nigel's Bright Waves – the future is uncertain. 'I'm sure that when this property is eventually sold,' she continues, 'the garden will be trashed and *tidied* – a word that I and most plant-lovers hate.'

To describe Sussex Emerald as 'gleaming green' is no hyperbole. Will and I are sitting with David Walker and Jac Turner-Moss as they count the moths in the Observatory traps, mugs warming the ground by their feet. Counting is no simple task this morning, for every egg-tray is laden with hundreds of sharply pointed, chestnut-hued micromoths. The local contingent of *Synaphe punctalis* evidently emerged during overnight warmth, for David estimates 15,000 of these gangly insects to be crammed into two traps the combined volume of a dressing-up box.

This leaves little room for other moths, and opening the trap prompts David to call for a gasmask as the winged dust-clouds billow outwards. Yet somehow there is still space for a single Sussex Emerald. Will and I gasp with delight.

Fresh as anything, this moth is every bit as gorgeous as we had been led to believe. Like waves, emeralds are geometers. The leading edge of a Sussex Emerald's wing is straight, but the trailing fringe of the hindwing undulates to a sharp point. The rear of each wing is bordered white and sporadically spotted with terracotta. Two white ribbons wave across peppermint forewings. Surprisingly pink legs and white antennae complete the look. Sussex Emerald is stunning.

It is not the only local speciality to find a berth in David's traps. Four White Spot evince the continuation of that rarity's flight season. A slender, insignificant-looking moth turns out to be a Pigmy Footman. The smallest of its lichen-munching genus, this fade-to-grey critter is nationally near threatened. Its tiny British range comprises east Kent – from where it was described as new to science in 1847 – and sandhills around Neil Bowman's cottage in north-east Norfolk. Not every rarity is a looker.

One that is, however, requires a third Dungeness visit, this time in early August. My girls have given up on me entirely, seeking solace in the Italian sun, a pasta-making course and a swimming pool. Their absence at least frees up the passenger seat for the Wingman. Swings and roundabouts.

We are after Scarce Chocolate-tip, a moth as delicious as it sounds. Britain's trio of chocolate-tips follow the same basic model. Roughly the length of the outermost joint of a little finger, they are shaped like a Buff-tip but enhanced by a hummocky thorax and, marvellously, a raised abdomen tip dipped in molten dark chocolate that protrudes upwards through closed wings like a happy dog's tail.

As its name suggests, Scarce Chocolate-tip is the rarest of the triumvirate. Britain's inaugural breeding colony was discovered by Henry Knaggs, who Michael Salmon describes in his book *The Aurelian Legacy* as 'a big, bluff, jovial man with a hearty, seemingly dashed-off style'. Periodically, at least, Scarce Chocolate-tip has resided at Dungeness since the mid-nineteenth century but may now be as common around Sandwich Bay as in this shingle heartland. It is no easy species to see, though. The moth's spring generation, Sean Clancy warned us, 'is as rare as hens' teeth', so we are targeting the year's second crop. Even so, David Walker has caught just five at the Observatory across thirty years. Our hopes are modest.

Will operates traps at Dungeness Long Pits, focusing on the waterside willows favoured by Scarce Chocolate-tip caterpillars – but without success. A mile away, I do the same at the RSPB reserve, where diving beetles clamber indefatigably towards my lights, drawn like a parched man to a desert waterhole.

The dark hours are bereft. So too dawn, when I examine my three traps. Nothing. I check the final egg-tray and tap out the remaining moths. Nothing. Failure.

I twirl the trap to check that no malingerer clasps its outside. And there, in the final ninety degrees of rotation, sits a Scarce Chocolate-tip. The moth had been there all along, unseen and somehow undisturbed by my banging, lifting and spinning. It remains tranquil and trusting as I encase it in a pot and phone Will.

The moth sits with tassled forelimbs gently extended. A KitKat bars the centre of its head and furry thorax. Two black beauty spots smile towards the back of its forewing, below which a Yorkie-toned wedge glints amber. And behind them all, the 'tail', standing to attention, with the actual chocolate tip. The moth quiescently waits out our admirations until I ease it along a willow branch, where it resumes unfazed languidity.

This winged confectionary is not the only beach-bum wonder in our traps. Pale Grass Eggar is a fur-ball with antennae as generous as a Highland cow's horns; and its straw-coloured local representatives are another Dungeness exclusive. The brick-flecked Rest Harrow is superficially similar to Bright Wave; it seems to be enjoying a range expansion, with Dungeness among the grateful beneficiaries.

Both species delight, but my eyes brighten most at Plumed Fan-foot, named after the feather dusters that serve as the male's antennae. Fan-foots more generally derive their name from fan-shaped appendages on males' hindlegs, which waft pheromones towards receptive females. Found new to Britain by local moth-er Barry Banson as recently as 1995, this curious insect has swiftly become resident along the Kent and Sussex coast. It has even been found in the garden of London's Natural History Museum, about as urban a location you can get. Previously known from Central Europe south, its establishment in Britain affirms the sense of an era as dynamic as the shingle on which we caught it.

Throughout this Year of Moth, I meet, chat or correspond with innumerable souls whose interest in moths dices with the obsessional. Several run a trap every night of the year, whatever the weather. Some readily travel hundreds of miles to 'pot-tick' a wind-blown waif.

Others dedicate long winter nights to gazing relentlessly down a microscope as they prise apart the wedding tackle of minute moths. One even appears to have no topic of conversation beyond moths.

As someone suffering from obsessional tendencies, I can empathise. Friends accuse me of an addictive personality. Fortunately, my malady now tends to exhibit benign symptoms – compulsively drinking herbal teas, binge-watching Scandi noirs and listening to songs of indie-rock legends The Wedding Present on repeat. But such expressions of devotion seem feeble beside that of Adrian Spalding, recent president of the British Entomological and Natural History Society. The intense fascination harboured by this moth statesman – a man of imposing mind with an authoritative manner – is not merely for moths in general, but for a single species above all: Sandhill Rustic. Spalding's four decades of dedicated study include eight consecutive years of near-nightly visits to a single isolated colony during the moth's late-summer flight period, plus trips to see the moth in five European countries. The outputs are magisterial: he has devoted both a website and a book to its life and times. Tongue in cheek, he confessed to the magazine *Atropos* that a mated female Sandhill Rustic would be his Desert Island Discs 'luxury item'.

You might think this madness. But Sandhill Rustic is bizarre enough to justify Spalding's biographical endeavours. The moth's British choice of landscape is oddball. Here it is exclusively coastal, a seaside existence placing it under existential threat from surging tides and rising sea levels. Yet across Europe, Sandhill Rustic routinely lives inland – and is even a mountaineer, reaching 1,600 metres' altitude in Andorra. Its British distribution is wacky: isolated populations of four forms taxonomically differentiated as subspecies, separated from one another by nearly 200 miles and each evolved such that their colour matches their home's substrate.

The subspecies that Spalding studies occurs nowhere else in the world but a single Cornish sandbar, where it occupies an area smaller than a typical cricket pitch. He learned that, on average, these moths move just eighty-six metres during their four-day adult existence. Females may not fly at all; one that Spalding watched sat motionless for nearly six hours at night, yet dropped to the ground at dawn before crawling 103 centimetres in nine minutes. Each year, he counted many more females than males. Such a gender-balance strategy is unusual among butterflies and moths but makes sense in cramped populations where males do not need to travel far to unearth females.

All this pales into insignificance, however, given Spalding's discovery that adult Sandhill Rustics are at home underwater – an element that sounds the death knell for almost all British moths. A century ago, two entomologists observed adult Sandhill Rustics emerging underwater from their pupae and crawling up inundated stems towards the air. This inspired Spalding to conduct an experiment, gauging the responses of adult moths to rising water levels in an aquarium. Half the moths tested actively moved down into the saline water, clinging to stems to avoid being washed away by the rising tide. One stayed underwater for a full hour, apparently unharmed and seemingly wrapped in a film of air. Most bizarrely of all, five even appeared to kick their legs and swim.

To see this barely credible creature, the Wingman, Dave Andrews and I head to Bawdsey Quay. To the wind-rushed clanging of yacht halyards and the gabbling tide, dusk sees us bumble across light dunes tethered by Marram Grass on to the saltmarsh that somehow, somewhere, secretes Suffolk's Sandhill Rustics. The subspecies here, *demuthi*, was first encountered along south-east England's muddy shores in 1963, yet misidentified for a decade as the widespread Flounced Rustic. Its larvae thrive in muddy, oxygen-starved saltmarsh beside tidal pools – a profoundly damper life than other British populations, which occupy sand or shingle.

We leap glutinous creeks whose salty liquid has been temporarily sucked away by the chestnut half-moon's gravitational pull. Locking the generator to a fence between electric-blue fuel containers marking the highest of tides, we fire up the lights in the knowledge that we will need all equipment off site before the sea reasserts itself an hour before dawn.

Night's curtain falls and, slipping around by torchlight, we spot several Saltmarsh Plume standing boldly atop the lilac inflorescence of Common Sea Lavender. These straw-coloured micromoths are poised like a catapult, wings raised at an acute angle, straining high above the head. Gingery, shaggy-maned Ground Lackeys slump around our traps like exhausted lumps of toffee. Seeing them here is a relief, since it avoids the need for us to search at a Kent nudist beach. But by the time we head for our car-beds at the end of the witching hour, this promising start has not translated into Sandhill Rustics.

We are fretful, but only marginally. Local expert Matthew Deans had warned us that Bawdsey's Sandhills fly very late at night.

Sure enough, when we wriggle out of our sleeping bags a scant two hours later, then trip through the pitch towards our lights, we smile. Three traps house eight somniferous Sandhill Rustics. The subspecies *demuthi* is known to be variable in appearance, and the octet make the most of their individuality by throwing every shape and tone – as long as it's essentially brown – at their mien. They exhibit dark blotches and pale stripes, stitches and sweeps, kidneys and ovals, edging and etching. The more I look, the more I see. And the more I see, the more I admire Adrian Spalding for his choice of muse.

14

Rock and a hard place

Somerset, Conwy and Dorset
June–July

Five hours after leaving home at dawn, Wingman Will and I are ascending a perilously smooth, daftly steep rock face on the Somerset side of the Avon Gorge with a woman whose daughter is also called Maya. It's an odd coincidence over which to bond but, in the absence of climbing ropes, I'll take anything that tethers me to Jen Nightingale. Alongside a day job running Bristol Zoological Society's UK conservation programmes, Jen somehow finds time to study for a doctorate and run annual surveys of one of Britain's rarest moths.

Silky Wave was first spotted alongside Avon Gorge in 1851; its English range also once encompassed pockets of Buckinghamshire, Dorset and Norfolk. Nowadays, this near-threatened moth occurs at just two other British sites, both in coastal Wales. It is hard enough to spot the dots on a distribution map, let alone connect them. In conservation terms, each population is trapped. But that makes saving them all the more important.

'Guided by Butterfly Conservation, I took responsibility for the formal surveys nine summers ago,' Jen says. 'Random counts weren't enough. We've standardised things, counting moths along transects at priority and satellite sites.' It sounds simple, but the reality isn't. Rock makes for a hard place to do conservation.

The Avon Gorge formed during the last ice age, when glaciers forced the eponymous river to excise a new route through carboniferous limestone. Today bare rock stands proud, far above the waterway. Scrubby slopes and dense woodland largely shroud three centuries' worth of mining and quarrying. It is through Leigh Woods – home of Britain's original but long-since-extinct population of Scarce Hook-tip – that we hurriedly descend to meet Jen and fellow ecologist Neil Green north of the 150-year-old Clifton Suspension Bridge.

On this warm and sunny morning, the pair have already surveyed two sites. 'Great news!' Jen exclaims. 'It's moth heaven out there. There's loads of Silky Waves flying. Last week, we caught the peak flight for the Bristol side of the Gorge. This week it's Somerset's turn!'

After a five-hour drive on the back of fitful sleep, this is welcome news. Our spirits lift. Then subside when Jen gestures towards the cliff that we must scale if we are to see Silky Wave.

Neil the human ibex glides up the steepest, barest slope without recourse to footholds or grabbing at vegetation. Jen leads the ascent of an adjacent scrubby, well-grassed flange. Jelly-legged, Will and I follow. Scrambling and slipping, we gradually gain altitude. On the odd occasion when we have mastered our wobble, we gently prod the dense, knee-high scrub. Although temporarily intrusive, this modest form of disturbance is the accepted survey technique to coax somnolent Silky Waves from concealment – the best way of counting them.

And there are plenty to count. As an adult Peregrine screams overhead, we tally seventy-plus Silky Waves across two quarries, taking Jen and Neil over 200 for today. They are ecstatic, both evidently caring deeply for this animal. This is no superficial love based on mere appearance, for Silky Wave is an unassuming-looking waif. Its scientific name, *dilutaria*, broadly means 'washed out', a reference to its insipid pallor and faint ripples crossing splayed ivory wings. But this moth has something. And it needs us – which, in the Avon Gorge, means it needs Jen.

'There's a massive problem with invasive cotoneaster and Holm Oak across two quarries where the moth lives,' says Jen. 'Fortunately, Butterfly Conservation has funds to do some eradication, and goats are grazing another quarry.' Jen's team must be doing something right because the population has inflated this century. Even better, this year proves one of the best since formal surveys began in 2011, with record counts at five of seven sites in North Somerset's portion of the Gorge. But Jen is concerned at declining numbers on the Bristol side, where something seems awry.

Determining precise population levels is tricky. In 2013, Jen trialled a technique known as 'capture–mark–recapture'. Moths are trapped and individually marked with colour. The proportion of marked to unmarked moths seen subsequently enables surveyors to calculate populations. 'We failed dismally!' Jen laughs. Her team tagged fifty-two Silky Waves but, over the following two days, only recaptured a handful – and instead counted 190 colour-free moths.

Uncertainty seems prevalent wherever rare moths are concerned. In *Moths*, E.B. Ford was perplexed as to why Silky Wave was not more widespread in Britain, given many locations nationwide could furnish the moth's habitat demands. In today's sole remaining English population, Jen is unclear about what sustains North Somerset's contingent, since the larva's supposed foodplant, Common Rock-rose, is sparse here. She is also unsettled by the moth's changing phenology. Its peak flight season used to be in early July but has advanced to the third week of June. Could climate change yet scupper Jen's decade of striving?

The circumstances surrounding the discovery of Silky Wave near the Avon Gorge in 1851 remain hazy. It appears to be John Sircom to whom hats should be doffed. Who Sircom was is unclear; he musters only the odd brief mention in scientific journals of the day – a blink-and-you'll-miss-him extra in the protracted dramaturgy of mothing.

For creatures so widely scorned today, the history of interest in moths is staggeringly long: in 400 BC, none other than Aristotle wrote about Wax Moth. Four centuries later Pliny the Elder turned his mind to moths. Curiosity was resuscitated in the sixteenth century as part of a wider interest in nature. The first British publication featuring the country's moths – Thomas Moffet's *Theatrum Insectorum* – appeared in 1634. In the seventeenth century, the Italian Francesco Redi and Dutchman Jan Swammerdam revealed the secrets of lepidopterans' life cycle, connecting caterpillars (then called 'worms') with the winged adults ('flies'). Swammerdam was later immortalised in the names of three micromoth genera.

British naturalists returned to the entomological fray during the eighteenth century. Eleazar Albin published *A Natural History of British Insects* in 1720, offering names for fifteen of the hundred moths he painstakingly illustrated. Two decades later, Benjamin Wilkes scattered his works with the first English names for moths, many of which survive: alongside tiger-moths and burnets are Peach Blossom and Golden Y, Gothic and Scallop Shell. This surge in ascribed identities came about because of the establishment, sometime between 1720 and 1742, of the world's first entomological organisation: The Society of Aurelians.

The title 'Aurelian' stems from the Latin *aureolus*, and apparently relates to the golden hue exhibited by the chrysalis of celebrated butterflies. The learned society's membership included Moses Harris, who would write and illustrate two influential publications, including the snappily titled *The Aurelian, or Natural History of English Insects: Namely, Moths and Butterflies. Together with the Plants on which they Feed*. Over and beyond moth names still in use today, we owe a debt to the Aurelians. Their illustrations encouraged familiarity with moths and butterflies – particularly once Harris had produced a popularised follow-up book. They described species new to science and started fathoming their ecology. They investigated and explained moth metamorphosis from egg to adult. They assembled the first collections of specimens, sharing them for collective appreciation.

During the nineteenth century, our understanding of moths deepened. One realisation was that species' genitalia differed such that each exclusive 'lock-and-key' arrangement largely prevented hybridisation. Techniques – such as sugaring, netting and searches for caterpillars – were developed that remain routinely deployed today. Entomologists began to capitalise on the unseemly attraction that moths have to light, taking lanterns into the wilds to catch the insects after dark – a prelude to mobile moth traps. Horizons broadened. The growing railway network permitted 'field clubs' with hundreds of members to travel widely through the British countryside. Select explorers roamed the world, returning with trays of foreign species. This opened up to fraud a burgeoning trade in rare moths: it was easy to swell profits by passing off a common European moth as British-caught, thereby raising its value.

In 1857 Henry Stainton was the first to integrate natural history with identification in books such as *A Manual of British Butterflies and Moths*. Stainton was perhaps the quintessential Victorian entomologist: a man of leisure and private fortune who considered the pursuit and dissemination of knowledge to be his calling. He founded and edited two seminal entomological journals (plus had a major hand in two more) and wrote a thirteen-volume, four-language natural history of micromoths. He held open house on Wednesday evenings when anyone who so wished could seek his advice, inspect his collections or peruse his library. Stainton levied no expectations of social status, class or gender on his visitors. Nor do there seem to have been any explicit or implicit barriers more widely among butterfly- and moth-hunters. Granted, in the 1870s, the Norfolk and Norwich Naturalists' Society counted three apiece of peers and baronets among its 135 members, but this seems to have been an exception. Studying nature, as Michael Salmon posits in his splendid book *The Aurelian Legacy*, 'transcended and dissolved' established social barriers such as class and income.

With their diligence and expertise, and inspired by the twin motives of learning and profit, Victorian lepidopterists became a force to be reckoned with. As the nineteenth century closed, Britain's moth fauna was very well known – almost comprehensively. Once you have put to one side recent colonists that weren't around to discover, there appear to be just four resident British macromoths that eluded the Victorians – all for understandable reasons. Slender Scotch Burnet is very locally distributed, while Small Dotted Footman and Fletcher's Pug are drab to the point of anonymity. The last – and the only one to evade my itinerary this year – is Silurian, which is still known only from remote Welsh moorlands.

Active throughout the mid-nineteenth century, Henry Doubleday was regarded by peers as the most knowledgeable lepidopterist of his day. Among the many new species for science he formally described and named is one of Britain's most mysterious moths, Ashworth's Rustic. And it is that creature – together with the equally anomalous Weaver's Wave – that leads me to North Wales in my personal homage to Victorian Aurelians.

After an interminable drive, it is 8 pm when we draw up at a rutted lay-by near Sychnant Pass. A ragbag assemblage of Choughs – aeronaut crows, big-dippering on frayed black wings – caw their name in welcome. Further west, the road freewheels down through a gilded, wooded valley with slate-roofed, white-washed houses before belly flopping in Conwy Bay. Above us, to the north-west, towers Conwy's mini-mountain of Allt Wen, which reaches 253 metres above sea level – most of them, we soon discover, from the road. A track undulates upwards, winding between crags of various greys towards a rounded slope that pits purple against the bottle-green of bracken and the occasional yellow froth of gorse. We imbibe the scene, unaware that a mere four nights later a nearby area will be set ablaze by arsonists.

Wingman Will, university academic Nick Watmough and I have volunteered our services for the mid-July 'Ashworth's Rustic weekend', an annual survey organised by Julian Thompson, the warden of Pensychnant Nature Reserve for three decades. By chance, we are joined by Robert Jaques and Back from the Brink's James Harding-Morris, whom Will and I met while looking for Welsh Clearwing in Sherwood Forest four weeks previously.

By equally happy coincidence, the Victorian moth-er who discovered Britain's first Welsh Clearwing near Llangollen in 1854 – John Ashworth – had also discovered the *world's* first Ashworth's Rustic at the same place in 1853. Henry Doubleday (he of 'sugaring' fame) formally added the wholly new moth to the scientific record in 1855. That year proved seminal in terms of revealing Welsh secrets, because celebrated entomologist Richard Weaver – who relished exploring remote regions of Britain – discovered what eventually became known as Weaver's Wave in North Wales, a moth that also had extended spells under the guise of 'Greening's Pug' and 'Capper's Acidalia'. It is heartening that this genuine discovery retains Weaver's name, for his previous claim to infamy – purporting to discover a strictly European butterfly, subsequently named Weaver's Fritillary, in the West Midlands – had rather sullied his reputation.

The year after Weaver's find, his Wave was encountered at Sychnant Pass. 'It is,' Julian Thompson says in an aptly craggy voice, 'the earliest known moth record for Caernarvonshire.' When Ashworth's Rustic was also found there in 1881, Sychnant Pass inked itself onto the moth-er's map. A gold rush of Victorian lepidopterists were soon holidaying in Conwy, seeking to add to their treasure troves by visiting Sychnant. They set a trend that remains today principally because, in Britain, neither species has been found outside a small, rock-rich sector of North Wales. Our presence tonight pays testament to the area's ongoing allure.

We have to put the effort in, mind. It may only be 500 metres from the car to the rocky moorland we decide to survey, but I have to scramble the steep climb through heather five times laden with gear before we are ready. Opinions oscillate about where precisely to position the eleven traps. I urge Will to move one trap that seems pointlessly dumped below a three-metre cliff adjacent to pasture where free-range chickens stroll. Grumbling in disagreement, he complies.

Seen from the road, our series of eleven bulbs 'looks like a little village', Julian says when he joins us late evening. The sight reminds him of a letter to a local newspaper, published shortly before King George VI's 1937 ascent to the throne, which clarified that the mysterious nocturnal lights hovering near Allt Wen belonged to moth-trappers. 'I find it remarkable that moths were so newsworthy when Britain was grasped by coronation fever,' Julian smiles.

We're fretting about coming all this way for nothing. The consensus among moth-ers familiar with Sychnant seems to be that we *should* catch Ashworth's Rustic but will need substantial luck for Weaver's Wave. The evening starts mild enough for James to remain in shorts, but largely clear skies and a full moon presage dwindling temperatures overnight. The breeze doesn't confer optimism either. Will puts on a brave face. 'Moths up here must be used to the cold,' he counsels. 'If they don't fly on nights like this, they're not going to come out at all.'

Will's right. As the light fails, moths are fizzing past us, teasing our clumsy net-swipes. Large Yellow Underwings are lapping wantonly at the pink, campanulate blooms of Bell Heather. It takes until 11 pm for night to assert its dominance, by which time we have separated, each ploughing a lonely furrow through thigh-high heather, torch waving hopefully through the peat-pool. Twenty-five minutes later, I pass the mini-cliff face where Will had wanted to place a trap. On its lichen-patterned rock a blindingly obvious Weaver's Wave is resting marginally above head height, erroneously assuming that the dots and contours on its outstretched triangular wings render it invisible.

Panicked, I dither, flipping between whether to take a photo or ease a pot over the moth. The flash could prompt the moth to flee, meaning nobody else would see it. But the same could happen if I try – and fail – to coax it into a pot. The first approach would at least mean the others would believe me, but they would also fume at my ineptitude. I plump for potting; the priority is to give everyone a chance to leap with joy. I succeed: 'Weaver's Wave! I have a Weaver's Wave!'

The group dribbles over in response to my hollering, excited but unable to rush across rough terrain. The near-threatened moth we needed fortune to catch is far prettier in real life than in its theoretical book form. Its slender wings grant a more lissom mien than a typical 'geometer' moth and it wears lace underwear. Frilliness oscillates across one wing and on to the other, tracing a route as up-and-down as the track leading from the road. Will frowns as he realises the Weaver's Wave might have entered his trap – had I not pressed him to relocate it.

We check the traps every half-hour, occasionally topping up the generator's fuel quotient. There are plenty of moths of high-altitude, rocky terrain – exciting stuff for moth-ers hailing from low, soft places. Crescent Dart is a speciality of cliffs and hard coasts along western Britain. Annulet is another rocker whose grizzled greyness merges with the surface of boulders. The slate-toned Northern Rustic looks hewn from granite. James and Robert get a particular kick from a dusky micro, *Matilella fusca*. James calls it the 'night's Emma Stone – the unexpected star that wasn't billed on the poster, thus making its cameo all the more pleasing'. We get bewildered by an unremarkable noctuid before realising with a giggle that it is a Confused. Even better, its neighbour in the trap is an Uncertain. Five sleep-hungry blokes roll around the heathery turf in hysterics.

The clock strikes two. Ashworth's Rustic remains an aspiration. I am mindful of Julian's admission that not all annual surveys succeed. Weary and perturbed, I retreat to our base camp – a flattish, turfed promontory. Rolling out a sleeping mat between the boulders, I slump inside my bivvy bag. The others persist until, one by one, exhaustion claims them too.

As ever, first light comes all too quickly. We stumble upwards and riffle through the traps. We are buoyed to discover three further Weaver's Waves and – in the final trap – John Ashworth's finest hour: a delicately blue-grey moth with long, sturdy wings that are Etch-A-Sketched with jet and suffused with smoke.

We bounce back to base camp to log and photograph the catch. The ground is strewn with expedition detritus. Generators and fuel cans mingle with sleeping bags and mats. Eleven moth traps pockmark a grass ridge sequined by scores of moth-pots. There are cable reels and camera bags, water bottles and half-empty packets of oatcakes. Although James and Robert's incessant banter is undeterred by fatigue, Nick looks broken. By 9 am, his Fitbit informs him he has had precisely zero hours' sleep that night and already completed his 10,000 steps for the day.

As we release the moths, we reflect on the bizarre distribution of both target species. Ashworth's Rustic and Weaver's Wave are most likely interglacial, bare rock-loving relics that weathered the last ice age by occupying milder, coastal land before moving upslope as the cold retreated. Either species may conceivably cling on somewhere else, as yet undiscovered, just as it took years to realise that Ashworth's Rustic occurred in parts of Europe rather than being a British exclusive. Recent data, however, hint that the distribution of Ashworth's Rustic might be shrinking.

This doesn't surprise Julian. 'In about 1907,' he says, 'Manchester Field Club came here and collected 106

Ashworth's Rustic caterpillars so they could breed them through to adulthood, then pin them in their collections. We've never found anything like such numbers. When we look for caterpillars, we only happen on the odd one. It would appear that the moth has declined. Either that or we're no good at searching.' Tonight, though, our questing has succeeded. We have paid our respects to yesteryear's Aurelians.

After a long drive back and a day's breather followed by another protracted car journey, Sharon, Maya and I install ourselves at the bird observatory on Dorset's Isle of Portland – for a family-holiday-cum-mothing-trip. Triangular both on a map and in cross-section, this massif protrudes five miles into the English Channel, dominating the coastline when viewed from east or west. The rocky island rises steeply from Chesil Beach – its umbilical cord to the mainland – rapidly reaching nearly 150 metres elevation before subsiding sleepily towards France.

Portland Stone – the island's off-white limestone bedrock – has been quarried since Roman times and in substantial demand since the fourteenth century. The east side of Buckingham Palace has twice been faced with it, while St Paul's Cathedral owes its essence to stone extracted from Portland's Broadcroft Quarry. It is here that I spend four humid hours pretending to be a Victorian moth-er, equipped only with net and (in lieu of paraffin lamp) headtorch.

The earliest Aurelian to explore the island was James Dale, who travelled Dorset on horseback, diligently recording all that he discovered. Once the island became accessible to visitors through the Portland Branch Railway, in 1865, entomologists thronged here. Their number included Nelson and Helen Richardson, later honoured through nomenclature for their role in revealing to the world the curious, rare micromoth *Eudarcia richardsoni* (Richardson's Case-bearer), which I saw here in January.

'What astonishes me is quite how good entomologists the Victorians were,' says Martin Cade, in his twenty-fifth year as bird observatory warden. Martin is likeable and laidback, his weather-wizened face freely breaking into a broad smile as he generously imparts his experience of moths and their history. 'Helen Richardson would be wearing bustle skirts as she and Nelson wandered the island by day and night. Their only light source was paraffin lamps. Yet the Victorians barely missed a single species, however obscure. They were bloody good.' Nelson Richardson, in particular, seems to have been an underrated scientific genius. He even found a new species of plesiosaur – subsequently named *Cryptoclidus richardsoni* – buried beneath his garden.

Broadcroft Quarry now sits cheek-by-jowl with a Butterfly Conservation reserve of seven hectares. Before twilight warps my perception, I am astounded by the richness of the herb layer straining between limestone chunks and assailed by the heady scent of Wild Thyme. But the surprise is nothing compared to my stupefaction once the moths reach for the darkening sky. I have never experienced anything like it. Broadcroft's nocturnal world is in constant motion. My torch beam snow-globes with moths. In human terms I am alone, yet thousands of moths keep me company. There are hundreds of *Oncocera semirubella*, a chalk-loving moth as pink and yellow as a Fruit Salad sweet; moth-ers nickname it 'Rhubarb and Custard'. Several flavours of wave include, thrillingly, Portland Ribbon Wave, a local speciality that temporarily enters my net before resolving to avoid confinement. There are three species of plume, Kent Black Arches and Rosy Footman. There are stunning rock specialists – Marbled Green and Brussels Lace – whose disruptive patterning would render them invisible on lichen-bedecked boulders. Fifty Brown-tail moths dance four metres up before thunking into a bramble. I see several Chalk Carpet, a scarce moth that has vanished from half its British range since 1990. Unknown larger moths scud overhead, while tiny ones tantalise then evaporate. There are moths everywhere. The experience leaves me overwrought.

I grant myself five hours' sleep – this is a holiday, after all – before Maya and I sift through the repurposed dustbin into which Martin rehouses the night's moths for visitors to peruse. Maya gawps at the cuddly furriness of an Oak Eggar, a chocolate orange half as big as a smartphone but with a ginger hipster beard. The moth has nothing to do with oaks, but its cocoon vaguely resembles an acorn. Maya then gasps as a Garden Tiger parts its cream-and-black forewings to reveal scarlet hindwings spotted with large ultramarine eyes that would make a tropical butterfly proud. Maya's lovestruck face turns to dismay as I recount this moth's catastrophic decline – an 88 per cent crash in abundance from 1970 to 2016. In July 2003, a single trap at Yorkshire's Spurn Point contained 1,683 Tigers, and Wingman Will recalls estimating 900 carpeting the lighthouse carport wall. Never shall we see such numbers again. Of the 500 records in Bedfordshire's moth database, the past decade accounts for just five. The decline is accompanied by dwindling genetic diversity and, oddly, by a subtle change in wing shape. We cheer up by playing a moth game: can Maya count more Lackeys than I can Crescent Darts?

Thus inspired, it is Maya who drags me to the rocky coast east of the bird observatory to search for two teeny-weeny Portland specialities. We belly along the low cliffs then gently rummage among upstanding platoons of Golden Samphire with their succulent longitudinal leaves and sunny flowers. We come across the sharp-fronted *Epischnia asteris* but are more taken with the perky, upstanding *Aristotelia brizella*, a Thrift-loving species. As the waves start whispering our name, Maya trumps these scarce micros by spotting the bonsai delight that is Small Marbled. The colour of a Rich Tea biscuit, this tiny macromoth is ostensibly a migrant to Britain, but Martin suspects they now breed on Portland. Maya frames her find in toymaker marketing terminology: it's either a 'limited edition' or a 'rare'.

Our island sojourn ends all too soon, just two mornings later. We pause our departure at the northernmost tip of Portland, by the junction with Chesil Beach. I cannot resist paying final respects to the Victorian entomologists who grasped the significance of this place. How they discovered *Scythris siccella*, a member of what even mini-moth gurus Phil Sterling and Mark Parsons characterise as 'a group of rather obscure micromoths', is beyond me. Adults are miniscule and fiendishly difficult to identify. Moreover, *siccella*'s sole

British site is a sandy strip roughly a hundred metres long and three wide. I stroll beside the shallow ditch and scrutinise the larval foodplant of Sheep's Sorrel in full awareness that I have no chance of finding the moth. This for the simplest and saddest of reasons: it is no more.

'*Siccella* hasn't been seen on Hamm Beach for several years,' Phil concedes, 'and we don't think it can occur there any more.' Despite conservationists' best efforts, the larval requirements of juxtaposed bare sand and young herbs have been usurped by colonising grasses and Black Mustard. 'And if it's gone from here, then that is it for this moth in the UK – extinct at its only known location,' Phil concludes. The Richardsons, with their bustle skirts and paraffin lamps, would be unimpressed at our guardianship of the Victorians' entomological legacy.

15
Heather

Dorset
July

Darkness is levitating from the heathery swell, its cloak measuredly erasing contours and colours, gradually thickening the air around me, eventually obfuscating the sky above. In human terms, I have Purbeck's Slepe Heath to myself. Beside me, a male Nightjar bosses dusk, pulsing an electronic trill from a pine tree before aching into the air, its obtuse angles banking around me in curiosity. Or perhaps animosity, if I have unwittingly transgressed into the bird's territory. Precautionarily, I shunt three moth traps a hundred metres upslope.

Wisely purchased by the National Trust five years previously, Slepe Heath is among the Dorset heathlands conjectured to have provided Thomas Hardy with literary inspiration for *Return of the Native*'s Egdon Heath. More assuredly, Slepe has proved a critical piece in a jigsaw now enabling a novel approach to conserving Dorset's remaining lowland heathland.

This landscape is our family's favourite. Days among the purple haze of ling, pottering to the tune of lazy chattering from unseen insects, are a summer must. Typically, we look for Dartford Warblers or Silver-studded Blue butterflies. Never have I dedicated time to heathland's particular moths. A July family holiday rectifies that. Half a dozen moths vie for my attention, but one audacious name stands out… because it is probably extinct in Britain. 'Come to Dorset and try for Speckled Footman! Rediscovering it would make quite the scoop!' suggests my childhood friend Durwyn Liley.

Durwyn and his wife, Sophie Lake, are verge-of-hippy ecologists who know as much as anybody about how southern England's heathlands function. Modest and unstintingly generous, Durwyn runs Footprint Ecology, an acclaimed, socially responsible consultancy. When I visit, Sophie is jointly managing Dorset's Heathland Heart, part of the Back from the Brink portfolio, which she sees as 'a chance to do something really practical to change the fortunes of declining heathland species'. For a good – if upsetting – reason, Sophie's project isn't targeting Speckled Footman, a dirty-white sweep of a moth that is dotted like a domino. Across its historical British range of Dorset and the New Forest, Speckled Footman has only been seen three times in a decade, all in one small portion of Purbeck, and not at all since 2014. Bookies would offer short odds on its national extinction. Its apparent disappearance is an indicator – an indictment, even – of the parlous condition of Britain's heathery lands. As Robert Macfarlane beseeches in *The Lost Words*, the book of spells that he and Jackie Morris produced to conjure nouns from the natural world that have vanished from a children's dictionary: 'Remember heather, the company it keeps / its treasure'.

Lowland heathland is marked by bewildering duality. On a clement day, to shamble amid Bell Heather and

Cross-leaved Heath, between coconut-scented gorse and birch, is to court tranquillity. To follow the same route in wind or rain is to endure hostility. One day you can lose yourself; the next, you can get lost. On an intense summer's afternoon, there can be a world of wildlife within a hundred square metres. A winter's morning, however, presents as a morgue. As Slepe Heath flows from Stoborough Heath into Hartland Moor and on to Studland and Arne, the sense of wilderness is writ large. Yet heathland is an essentially anthropogenic landscape, one that extends beyond isolated pockets within forest only because prehistoric Britons razed the intervening trees. The duality also roams through our attitudes. Until the 1980s (and sometimes still today), heathland was synonymous with wasteland, ripe for urban sprawl or worse. Internationally speaking, however, it may be Britain's most valuable wildlife habitat.

Worldwide, lowland heathland exists solely in Atlantic Europe, from Norway to Portugal. Britain is blessed with one-fifth of this heritage but we have proved inept guardians. Of heathland that swathed England in 1800, we have eradicated an astonishing 85 per cent. We have ploughed it up, supplanted it with darkly uniform trees, drowned it in concrete. Remaining fragments are in worse condition than any other British habitat. Heathlands' structural intricacy is levelled out, prized vegetation stifled by marauding pines and bracken. Friable, sandy soils are compacted by trampling feet and saturated with unwanted nutrients from domestic-animal 'fertiliser'. Wildlife is hassled by off-lead dogs or predated by housing-estate cats.

One surprising culprit is the cessation of traditional human use. Long-established management practices such as sand-extraction and heather-harvesting created microhabitats such as tramped trackways, shallow pools and bare sand that are vital for specialist invertebrates and plants. The Back from the Brink project aspires to change all that, reversing declines experienced by 60 per cent of Dorset's key heathland species. It is a bold, much-needed initiative. Back from the Brink is not fighting the battle single-handedly, though. A partnership of conservation bodies strides across the Isle of Purbeck: the RSPB, National Trust, Dorset Wildlife Trust and Amphibian and Reptile Conservation all run reserves here. Since 2012, the partnership has restored or recreated almost 500 hectares of heathland – an area equivalent to London's Olympic Park. Most excitingly, these initiatives united in March 2020, when Natural England launched the new Purbeck Heaths National Nature Reserve. This has linked 3,400 hectares of heathland that were previously spread across seven different management regimes – and placed them under a single guiding vision. Protecting an area the size of York elicited media pronouncements that this was Britain's first 'super' reserve.

Such landscape-scale thinking is complemented by an appetite to understand what wildlife survives on Purbeck's heaths. Former nature-reserve wardens Chris Thain and Abby Gibbs run the Moths of Poole Harbour initiative. Their multi-year survey of the area's under-appreciated insects sits alongside, and is funded by, the charity Birds of Poole Harbour. Chris and Abby's basic goal is to understand where moths reside around this

Dorset coastal waterscape. In 2018, they devoted 126 nights to surveying the area. They join a long line of Purbeck naturalists. In the mid-2010s, the Heritage Lottery Fund financed an ambitious citizen-science project retracing ecological surveys by conservation pioneer Cyril Diver – an unsung naturalist who led the UK's first statutory wildlife body. In the 1930s, Diver feared the burial of Studland's heathland beneath new houses. He marshalled 158 volunteers – including renowned experts of the day – into meticulously documenting its wildlife. Through listing 2,500 species, they demonstrated the peninsula's hitherto-unrealised importance and rolled the pitch for Purbeck's protection.

Eighty years later, surveyors enumerated 3,800 species, more than half again of Diver's total. The respective aggregates speak more of available research capability than an improving environment, however. Modern surveys nearly doubled Diver's moth count, but they failed to find one-third of his assemblage. More than a hundred species of moth have simply vanished from Studland. Comparing the two surveys reveals widespread underlying ecological impoverishment. In Diver's day, heathland was open, connected, grazed and virtually tree-free. Today it is enclosed, pine-contaminated, scrub-dominated and largely without the pockets of bare ground on which the most vulnerable creatures thrive.

Speckled Footman may be the most troubled of the lot – if it's not already too late, that is. Although always sparsely distributed, it has been locally numerous. In 1890, Victorian entomologist Eustace Bankes recounted visiting Ringwood in east Dorset, 'bagging 18 ... by dint of really hard work in the broiling sun'. Celebrated moth-er and field-guide author Paul Waring has written that it was 'previously not uncommon to record 40–50 in a night without difficulty'. This century pens a different story. The last person to see a British-bred Speckled Footman was fellow field-guide author Chris Manley. 'When I discovered it on the Trigon estate twenty years ago it was very scarce,' he says. 'But I did find a small clearing where I could catch four or five per night. I even spotted one on my front door.' Since 2014, however: zilch.

Refinding Speckled Footman will involve leaps of faith – partly because its needs elude our understanding. Some moth-ers moot a fine-leaved, densely tufted grass, Bristle Bent, as a key larval foodplant. But this grass is common at Slepe Heath, where Chris and Abby have surveyed for twenty nights during the moth's flight season without a sniff. Other moth-ers berate excessive grazing or the incursion of pine trees and scrub as woodland seeks to reclaim former terrain. Meanwhile, former RSPB warden Bryan Pickess recalls a night when visiting moth-ers lined Soldier's Road, which connects three key heaths, with a dozen traps. One inference is that – in this part of Purbeck, at least – collectors may have hastened the decline of an already dwindling population. None of this gives conservationists anything substantive to go on. 'The trouble is, nobody knows,' Manley concludes. 'You can have all sorts of ideas, but there's no way of testing them.' Metaphorically as well as literally, we are in the dark.

Such uncertainty doesn't faze Durwyn. For months, he has been forensically planning a grand survey for Speckled Footman in Wareham Forest. The moth was recorded at two spots here in 2010 and at others previously. The Forest also abuts Trigon, location of the last-known record. Durwyn and his colleagues Zoe Caals, Chris Panter and Phil Saunders have searched the Forest for Bristle Bent. If Speckled Footman has a thing for this grass, there's got to be a chance that it hangs on near one of the grid references in Durwyn's resulting gazetteer. All that is needed is a bevy of volunteers to run traps.

Hence Durwyn's invitation to me. Hence our family holiday in Purbeck, scheduled for the flight season of Speckled Footman.

Durwyn and I bookend the Wareham Forest gambit with nights mothing on Stoborough and Slepe heaths – a warm-up followed by a warm-down. The sites are merely a mile apart, yet the experiences feel very different. At Stoborough, I find it hard to settle. It is carnival weekend in nearby Wareham; between exploding fireworks, a band blares Duran Duran covers, shredding the crepuscular calm. Cattle mooch about, risking being dazzled by the moth-trap lights. Roosting Silver-studded Blues sulk at my torchlight. It is well into the night before Durwyn turns up; until then, my loneliness puts me on edge. Worse, the cool night cramps moth activity. We call it quits in the early hours.

Slepe feels contrastingly munificent. Some locals have gathered to admire Nightjars; they gossip with Durwyn as I position traps around abundant Bristle Bent. Coming mid-heatwave, the night is beyond balmy. Once Nightjars have departed to hunt, we ride the warmth, exploring the heather and illuminating marshy tussocks. Bryan Pickess caught Speckled Footman here in 1980 before conifers spent three decades invading the heathland. Those trees have very largely been removed, so we harbour hope of a spectacular rediscovery. We call a temporary halt after midnight, treating ourselves to four hours' kip. But I'm too pumped to slumber and so return ahead of the rising sun. A Sika Deer whistles an alert to Durwyn's arrival, then we examine the catch as a Mediterranean breeze ushers in what will be Britain's hottest ever day.

Neither night delivers Speckled Footman, but both offer insights into heathland mothdom. True Lover's Knot and Beautiful Yellow Underwing prove pleasingly common doppelgängers. Both are variegated, swirled and striped with auburn, black and white – disruptive camouflage for secreting themselves in the bitty shadow of a scouring pad of Bell Heather, with its pendulous, deep pink blooms. True Lover's Knot derives its romantic name from white loops twisting across its wings; one tug would entwine the lovers. Beautiful Yellow Underwing is named for its unexpected gold-bling hindwings, perhaps flashed to momentarily stun a predator who has seen through the upperwings' bark-like disguise.

Two subtle, scarcer moths are localised specialists of southern England's lowland heathland. It takes practice to determine the neat black stitching along the trailing edge of the wing that differentiates Dotted Border Wave from lookalike members of its genus. Confidently identifying Small Grass Emerald involves tracing the white contours that flow across its pale green wings. I catch one fluttering gently over frothy Ling, a tiny phantom before the headlamp.

The warmth of daytime invites exploration too. Lowens join forces with Lakes and Lileys. Sophie indulges us by showing spots on Slepe and Arne where Back from the Brink is recreating the old ways. Enchanted, we watch some diurnal stars of Dorset's Heathland Heart. A female Sand Lizard lounging. Mottled Bee-flies fizzing. A Heath Tiger Beetle basking indolently before scurrying off at a Cheetah's lick. Most special of all, we spend an hour watching Purbeck Mason Wasps – one of Europe's rarest insects.

This creature is nothing like the familiar garden-party annoyance. Crisply attired in chestnut, ebony and cream, this solitary wasp excavates burrows in open, sandy ground suffused with clay. Here the female raises her brood, nourishing them exclusively with larvae of *Acleris hyemana* micromoths. The wasp extracts caterpillars from webs on the tips of heather, stings them into paralysis, then helicopters them alive to her lair excavated from bare, sandy clay. Only the freshest food will do for her offspring. We watch one female return, small green caterpillar clenched between legs, then wriggle underground. It is the only *hyemana* I see all year.

No wasps of any description attend the early evening garden party that launches Footprint's Footman night in Wareham Forest. Nearly thirty people aged between seven and seventy-odd are munching sausages and potato salad. We catch up with old friends and meet new ones.

Chat turns to moths and Purbeck, Brexit and Bristle Bent, and anything else that strikes communal chords. Our number includes professional ecologists from the public, private and charitable sectors, but everyone is here in their personal capacity, all willingly sacrificing sleep for the chance to show that Speckled Footman has not vanished into the eternal night of extinction.

Maya scampers around with Simon Curson's son. Chris Thain and Abby Gibbs are here. So too is an Australia-based friend, Guy Dutson, coincidentally back in Blighty; although no moth-er, he needs little persuasion to join us. Footprint Ecology staff and partners are present in volume. Jackie Underhill-Day, widow of Durwyn's late business partner John, has brought her husband's moth collection: creatures killed in Rothamsted Insect Survey traps but reverently preserved on index cards to generate value in their afterlife. Flicking through them reveals a Speckled Footman, wings splayed. Is this a good omen or bad?

Our host calls for attention and we fall silent, attentive. 'Time is ticking on and those people staying for the night's mothing have a journey ahead to set up traps.' Durwyn suggests we splinter into seven teams to target different areas. 'There's not many moth records across Wareham Forest, so it will be fascinating to see what we catch,' Durwyn says. 'We'll regroup at 7 am to share the most exciting moths over croissants and coffee. Chris Manley is trapping at Trigon and will join us tomorrow.'

As I drive north-west along forestry tracks to where Speckled Footman was trapped in 2010, 'tomorrow' feels a long time ahead. Guy and Peter Moore, a chirpy Wareham birdwatcher, help link lights through patches of Bristle Bent in otherwise unpromisingly scrappy heather. The effort embalms in sweat. Five minutes' stroll south, Wingman Will calls upon Phil and Poole-based moth-er Marcus Lawson to run traps amid regenerating, sandy-banked heathland. It's nuts that Speckled Footman has clung on here more recently than the core areas of the new National Nature Reserve.

Nightjars bat around us languorously as we open generator chokes and swish nets. Every step – every moth – offers potential. Guy delights in his first Buff-tip since childhood. Dark Tussocks slob, forelimbs extended like a praying mantis. Dotted Border Wave proves frequent, but remains subtle. On the home run to midnight, Marcus discovers an Angle-striped Sallow, its custard tones undulating with rhubarb threads. It transpires to be only the eighth ever in Britain's most moth-rich county. I pounce upon a Scarce Light Plume, a tiny, gingery micromoth-cum-glider that rests with elongated wings perpendicular to its body. Although a Mediterranean vagrant, it is not wholly unexpected as this summer has already witnessed a record influx.

We check every pale moth we see, recording six species of footman in good numbers, but not Speckled. We are undeterred, for our efforts feel part of something mightier. If we don't catch the target, perhaps another team will. Around 2 am, the air chills, becoming crowded with unseasonal mist that thwarts torches and dampens moth activity. We take the cue and collapse into a stuttering, car-seat sleep. When dawn whispers, we rise and count every individual of every species. It is only when we regroup at Footprint Ecology's office that we comprehend the scale of the achievement. Between us, we have identified 3,867 moths of 204 species. 'Overwhelming,' says Will, speaking for us all.

Collectively, we admire the specialities while refuelling on caffeine, sugar and left-over bangers. There are a fistful of interesting micromoths with complicated names. One is the delicate, flaxen *Anania verbascalis*, a locally distributed denizen of southern heathlands. Upon release, the moth proves her gender by flickering intently towards some Wood Sage, where she immediately lays the next generation.

Two of Britain's titchiest macromoths also take the plaudits despite being so small they could be dismissed as micros. The Rosy Marbled is faded, but delicate pinkness and strongly marked rear wings betray its identity. This moth of sandy ground is restricted to southern and eastern England and has only recently spread west into east Dorset. This individual is likely the descendent of a recent colonist. Late July is tardy for this moth nowadays, mild springs accelerating its emergence compared to the 1970s. The other micro-macro is Marsh Oblique-barred, a nondescript snout-like moth whose scattered distribution across Britain has surged eleven-fold since 1990. Nobody knows why it is doing so well.

Even more pleasing are two Dingy Mocha, a nationally near-threatened moth that breeds only in the New Forest, Dorset and perhaps Devon. Grizzled grey but suffused chestnut, this butterfly-shaped moth worries conservationists because of its finicky ecological niche. Dingy Mocha not only lives solely on damp heathland and sandy mires, but lays its eggs uniquely on young sallow bushes growing in open conditions. On heathland, sallows are typically considered invasive scrub and thus, unfortunately, often cleared by well-meaning conservationists or browsed by cattle. As with Speckled Footman and Purbeck Mason Wasp, the narrow ecological requirements of Dingy Mocha do not make for straightforward conservation.

Dingy Mocha turns out to be the night's rarest moth. When Durwyn turns over the final egg-tray, no Speckled Footman is sat there. We slump despondently, dreams dashed. We have failed to rediscover this mysterious moth. All Durwyn's effort, in vain.

But there's always next year. By then, Purbeck's new, inspiring National Nature Reserve will be up and running – a suite of protected areas managed from the same hymn sheet to safeguard Dorset's heather treasure. And where there's hope…

Postscript*: less than ten months after our visit, the area of Wareham Forest where we had focused so much attention went up in flames, a discarded portable BBQ or campfire causing 220 hectares of ecologically valuable heathland to be transformed into ash. Disaster. Five weeks later, Durwyn Liley opened a Wareham Forest moth trap to find a Speckled Footman sitting on the uppermost egg-tray. Jubilation. Over the following two summers, Durwyn and volunteer surveyors find dozens. A moth feared extinct lives on. There is hope yet.*

16

New arrivals, welcome?

Kent, Bedfordshire, Norfolk, Hertfordshire and Greater London July–August

Jesus set the tone during the Sermon on the Mount: 'Lay not up for yourselves treasures upon earth, where moth and rust doth corrupt…' (*Matthew* 6:19). The indicted insects were clothes, carpet or meal moths – creatures whose larvae, even then, were infamous for their taste for human property. Attitudes to moths had not mellowed by the fifteenth century, when the Old Norse word *motti* (from which 'moth' is derived) was used for the caterpillars of clothes moths.

Admitting a fondness for moths risks making you a social pariah. For moths eat our textiles, flour and grain; they destroy our gardens, shred wild plants, cause our hospitalisation and even our death. Tabloids and broadsheets alike write apocalyptically of 'plagues' and 'infestations'. Opening a haberdashery chain catalogue, I confront a double-page splash of devices to terminate clothes and carpet moths. I am invited to purchase traps and sprays, hangers and sachets, cupboard-fresheners, moth-proof plastic bags in which to store clothes safely. In May 2019, a YouGov poll for Butterfly Conservation found that three-quarters of the public think negatively of moths. A sixth considered moths ugly; one in eight was scared of them. We hate moths and want them expunged from our homes and lives.

For sure, some charges are valid: given the opacity of the smoke, there must be some fire. Agriculturalists despise the larvae of Plum, Leek, Meal, Corn, Cabbage and Turnip Moths – the latter 'to the farmer the most serious pest', according to E.B. Ford – which can consume crops. Common and Case-bearing Clothes-moth caterpillars are undeniably guilty. Brown-tail and Oak Processionary Moth larvae can certainly irritate bare skin and airways.

But tarring all moths with the same brush? That rankles. The wrongs of a small group – just two types of British moth eat clothes, barely 3 per cent imperil plants – are extended to all 2,500 species. If one schoolboy is naughty, the whole school doesn't merit detention. Yet for the uninitiated, the unaware, the unthinking, all moths are bad. And bad to the exclusion of all of their goodness – their wonder, their beauty, their sensitivity to environmental change, their pollination services. Just one in five people interviewed by YouGov thought moths 'important'; not even a third considered them 'interesting'.

In this context, spending my year exclusively on 'good', 'important' or 'interesting' moths would be a cop-out. I must explore the dark side of moths too. All the more so because our globalised world constantly ups the ante. Just as air travel whipped Covid-19 so devastatingly between continents during 2020, so shipments of plants across national borders inadvertently import stowaway caterpillars, eggs or pupae – 'adventives' in entomological parlance. Meanwhile, climate change is enabling European species – natural immigrants – to move north and establish themselves in the British Isles.

Butterfly Conservation's Mark Parsons routinely scours entomological literature to track moth species new to the UK, summarising his findings in *The Entomologist's Record and Journal of Variation*. Since 2010, he has identified sixty such newbies. Many are one-offs or occasional visitors. Of the twenty-two that he judges likely to have colonised during the past decade, fifteen did so naturally and seven were 'probably human-assisted, for example through the horticultural trade'.

The pattern is similar across Europe. In 2010, Carlos Lopez-Vaamonde and colleagues logged eighty-eight European moths and butterflies that were expanding their range and ninety-seven non-native species that had established themselves in Europe – 1 per cent of the region's total. Of the new colonists, three-quarters arrived during the twentieth century. Since 2000, the rate has jumped to two new species per year. As befits transport predominantly through the globalised horticultural trade, more than half of the 'aliens' are confined to parks and gardens – the end destination of foreign plants seeking to sate our desire for the exotic. We reap what we sow.

Many new arrivals pose no harm to native species: 'new arrivals, welcome'. Others, however, are pestilent, causing considerable damage: 'a plague on both your houses'.

It is time to spread the net.

Two pots, two fridges, two kitchens. One in Kent, the other Bedfordshire. Two little moths: one new to the

Northern Hemisphere, the other entirely new to science. One Kiwi, the other perhaps Indian – but nobody really has a clue. Neither should be here, yet both are. Causing no problems, they are welcome.

Beneath the book jacket of its nondescript, monochrome appearance, *Prays peregrina* relates a remarkable tale. In autumn 2005, a mystery micromoth was caught in west London. By 2007, ten capital mothers had trapped it. But nobody could name it – not even moth boffins ('moffins'?) worldwide. The vast specimen collection at London's Natural History Museum offered no matches. By scrutinising dead moths' sex organs through a microscope, the Museum's David Agassiz narrowed it down to the genus *Prays* but could get no further because there was nowhere further to go. It was not merely unidentified, but unidentifiable. Living quietly and apparently exclusively amid one of the world's most vibrant capital cities, this moth was entirely new for science. The discovery was astonishing. We think of new species hailing from tropical rainforests, not urban jungles.

But there was a catch. Judging from the minutiae of its wedding tackle, the puzzle's nearest relative seemed to be a species from northern India and Nepal, *Prays curulis*. If that were the case, this conundrum probably originated from nearby. Although known only from London, this moth presumably wasn't actually British but most likely Asian. The country had another new adventive, which Agassiz reflected by naming it *peregrina* in honour of its peregrinations. He believed it most likely that the larvae or pupae were imported in an Asian foodstuff. 'But this is purely conjecture.' How could I not try and see such an unfathomable creature?

As I grapple with how, when and where to catch *Prays peregrina*, Kent moth-er Nigel Jarman comes to the rescue. Whether through repeated imports or an expanding population, the moth has reached Nigel's garden in Kingsdown. Fortuitously, that very day we are booked in for an afternoon cuppa. Our host taps the moth onto a gnarled chunk of wood. The size of a grain of rice, *peregrina*'s short, tented wings narrow to a point. Its unremarkable greyness is fettered only by a black heart on its nape and two sooty bands where its wings meet. For sure, it's not much to look at. But how little that matters given its unsurpassable back story.

Ten days later, another kitchen and another moth. Conservationist Richard Bashford welcomes Nick Watmough and me into his home in suburban Cambridgeshire (politically, at least; for moth purposes, confusingly, the Bashfords inhabit Bedfordshire). Richard's teenage daughter Molly is cooking a vegan dinner ahead of a planning meeting for the imminent Latitude festival. She seems unperturbed at the intrusion on her space.

To the accompaniment of chattering sparrows, Richard proudly shows us a *Trachypepla contritella* that he caught after last orders the previous evening. The micromoth is roughly the size and shape of Brown House Moth but much smarter. Black bars wave across its wings, isolating a broad pale splash. Behind this, oddly, sit two little tufts that stretch skywards. Such raised scales are unusual among British moths, and so piqued Richard's interest when he caught his first in June 2018. 'I had no idea what it was – it looked like nothing in the book.' Richard phoned a friend, Bedfordshire moth-recorder

Andy Banthorpe. Equally bemused, Andy did likewise, engaging Martin Honey at the Natural History Museum. Enthusiast Steve Nash was the first to suggest it might be a species from New Zealand. And so it transpired. Although widespread and fairly common in its native country, *contritella* had never previously been recorded north of the Equator.

'Just before Andy published the discovery in the scientific press, we learned that another local moth-er had caught this mystery species six years earlier,' Richard recalls, but the photos had languished unidentified on his hard drive. 'Fortunately, Andy managed to update the paper before publication. Then the moth recorder for Huntingdonshire started catching them too. Both gardens are within a kilometre of mine.' None of the three garden moth-ers had planted vegetation native to New Zealand – but someone nearby presumably had, for the chances of *contritella* dispersing naturally from the other side of the world are zero. In the absence of a formal English name, Richard grins, 'we call it the "Kiwi Enigma".'

If other people's gardens have such wacky moths, I wonder, might mine? I need not look far for the first interloper. My most regular garden moth this year is originally an Aussie, Light Brown Apple Moth. This fingernail-shaped tortrix comes garbed in fifty shades of brown and even more patterns besides. First recorded in Cornwall in 1936, it spread slowly until the 1990s before exploding in abundance and distribution. Writing in *Atropos* magazine, Jim Porter recalled that 'in early 1997, I had never seen the species, but by 2000 I could guarantee it appearing at light every night between May and September'. In Norwich, it is now so well established that it is the sole moth I record in every month.

Cornwall has form with the initial British records of a suite of Australian invertebrates. Also writing in *Atropos*, Malcolm Lee points the finger at the importation of Australian plants by Treseder's Nursery at Truro. It was, Lee observed, 'the premier Cornish nursery from Victorian times until closure in the 1980s'. Onwards shipments catapulted Light Brown Apple Moth into garden centres across the country and hence into private gardens. Even now, it is largely an urban and suburban species, less readily found in the wider countryside. For this reason, perhaps, early fears that moth larvae would devastate commercial fruit have not materialised. Fortunately, the residency of 'LBAM', as moth-ers abbreviate this creature, seems anodyne – so its right to remain goes unchallenged.

The same cannot be said of a more dramatic micromoth that is enraging gardeners and worrying conservationists. I first clapped eyes on *Cydalima perspectalis* (Box-tree Moth) in 2016 but had no clue what it was. This garden arrival was stunningly distinctive – the size of a Speckled Wood butterfly but pearly-white with a black felt-tip border, the composite form shimmying magenta and peach – but unequivocally absent from any field guide. It spoke of the tropics, reminding me more of a heliconia butterfly than a garden moth.

During 2019, I catch a handful most summer and autumn nights. Each and every one mesmerises but simultaneously disturbs. Originating from east Asia, Box-tree Moth was unwittingly imported into Europe with shipments of Box plants in 2007. The first British record came that same year, being mooted as a migrant. For six years it spread little here, but successive importations and domestic movements of 'infected' plants expanded its range from 2013 onwards. In 2015, a Royal Horticultural Society hotline received 383 reports of the species' presence; in 2018, the figure topped 6,000. It now resides throughout England bar the far north-west and has footholds in parts of Wales and south-east Scotland. Where caterpillar densities are high, it disfigures and even defoliates cultivated Box topiary. This winged messenger from the anti-globalisation movement punishes precisely the people responsible for its invasion.

Sadly, however, Box-tree Moth spares not the innocent. In Germany, larvae took just two years to defoliate 90 per cent of the country's largest native Box forest. Two years later, caterpillars munched their way through the bark, killing many trees. In the gaps created, faster-growing plants took root, permanently altering woodland ecology. Such form set British conservationists fretting because natural Box communities are scarce enough here for extirpation to be conceivable. In September 2018, the inevitable happened. Larva were discovered feeding on native Box at the well-named Box Hill on Surrey's chalk downland. 'Overall,' admit eminent lepidopterist Colin Plant and colleagues in a 2019 paper, 'we cannot escape reaching the conclusion that *C. perspectalis* must be regarded as a pest species.'

Box-tree Moth is not my only garden moth to have garnered a bad reputation. Persil-white with the tip of its upturned abdomen dipped in chocolate, an adult Brown-tail cutely professes innocence. Its caterpillars, however, have long been considered the spawn of the devil. In 1780, an infestation of communal webs of stinging larva greatly alarmed Londoners. Anyone removing them risked the caterpillar's hairs irritating their skin. So it continues today: in 2019, uproar at larval webs compelled a Hampshire school to evacuate its playground, a Norfolk council to destroy path-side webs and tabloids to express outrage and spread fear ('terrifying webs crawling with ugly bugs'). Hysterical, for sure, but not without cause. 'A lad helping us out one year had a terrible reaction when he cleared some web-covered vegetation,' recalls Martin Cade, warden of Portland Bird Observatory. 'He became blotchy and swollen. I had to take him to the urgent-care centre, where they gave him an injection. And it's not just the caterpillars. I gather that the adult's abdomen tip has similar irritating powers.'

I take greater care when releasing this summer's Brown-tails from my traps.

Mid-August. A long family weekend on our former London stomping grounds offers an opportunity to catch up with grandparents and friends – and to pursue further new kids on Britain's moth block.

We pause the M11 crawl by detouring to Hertfordshire's Rye Meads RSPB reserve. Birdwatchers love the site for its lagoons, botanists for its ancient flood meadows. With half an hour to locate my quarry while Maya and Sharon pond-dip, I make for neither habitat, beavering instead towards rank, thorny scrub dominated by bramble and wild-growing raspberry. Lest I were uncertain, a vibrant display announces this to be the place to see Raspberry Clearwing, the only recent addition to Britain's list of splendid wasp imposters.

'Raspberry patch... galls... stem... larva,' shouts one set of white-lettered, blue-painted signs. 'Raspberry Clearwing... pheromone lure... July–August... day-flying moth,' explains another.

In autumn 2007, local moth-er Jim Reid was pruning garden foliage when he noticed strange holes in some raspberry canes. Determining that they were made by Raspberry Clearwing larvae, he checked nearby fruit farms – where he found adults. Britain had a new species of wasp-mimic. Raspberry Clearwing is common enough in mainland Europe to be a plausible natural immigrant. However, the combination of circumscribed core range inland and paucity of coastal records suggests it reached our shores through the importation of commercial plants, in this case raspberry.

Despite limited time at Rye Meads, I elect not to 'cheat' by using a pheromone lure to attract the moth. Given that fruit-growers use the devices to rid crops of these 'pests', the irony of doing so would have been too bitter. As luck would have it, I quickly spot an adult Raspberry Clearwing sunning itself on a namesake leaf. It is a massy clearwing, longer and stockier than the similar-looking Yellow-legged or Six-belted, with narrow, densely packed stripes.

Jogging back to the pond-dipping site, I gasp under the line seconds before the agreed deadline. As Maya packs away her rod, I flush a chunky beige moth that conveniently perches low in a sapling. I am surprised to discern a male Gypsy Moth – another creature with a richly controversial history. Begging a minute's extension, I raise Maya onto my shoulders so she can observe his gingery mane, fractal-rich wings recalling its close relative, Black Arches, and deeply pectinated antennae. 'Look at his eyebrows!' Maya screams excitedly, if erroneously.

In Britain, Gypsy Moth – the creature that inspired aircraft-pioneer and surreptitious moth-er Geoffrey de Havilland's DH60 biplane – blurs the boundary between indigenous and adventive. Until 1907, it bred in East Anglian fenlands. The drainage of that once-mighty region plus the moth's restricted diet of bog myrtle and creeping willow put paid to its existence. Another creature banished… but not for ever. In 1995, a small colony of Gypsy Moth was discovered in north-east London. Located far from the native moth's Fenland foodplants, the population was clearly not an indigenous remnant but arrivals from Europe. Given that females are largely flightless – and despite evidence that the wind can disperse caterpillars significant distances – entomologists generally reckon the eggs that engendered this colony were transported to Britain via vehicle, possibly in imported timber. Populations have subsequently appeared from Dorset to Essex – and it even returned to its former Fenland range in 2018 – but Britain's cooler summers are likely to prohibit widespread expansion.

Just as well. In southern Europe, large numbers of Gypsy Moth caterpillars regularly defoliate broad-leaved woodland. Numbers are controlled by a predatory beetle, *Calosoma sycophanta* (the marvellously named Forest Caterpillar Hunter), such that the two species' populations form mirror-image parabolas over time. Britain lacks this particular Gypsy-muncher, so if the moth's populations were to take hold we would lack the checks needed to effect balance.

Looking across the Pond evinces the point. In North America, taxonomic naivety has sparked ecological devastation. In the 1860s, Gypsy Moth was erroneously judged to be a member of the silk-moth genus *Bombyx*. They are no such thing. Not only did the Gypsies brought to Massachusetts by French entomologist and artist Léopold Trouvelot not produce silk but, in 1868 or 1869, they escaped. Without specialised predators, Gypsies established themselves in the wild and subsequently ran riot. Despite eradication attempts, larvae have damaged swathes of forest across four hundred thousand square miles of nineteen US states. This granted the species inclusion in the world's hundred 'worst invasive alien species', as agreed by International Union for the Conservation of Nature experts. It is the only moth in the rogue's gallery.

It's probably best not to tell my mother that she has Gypsy Moths living in her garden – something I discover by running a portable moth trap under an apple tree behind their Hither Green home. Over two nights, Gypsy Moth features among a quartet of tales about arrivals and adventives that are generated by the small black cube I plonk on my parents' garden bench.

First out of the trap is an unequivocal good-news story. Eighty years ago, Britain's first Toadflax Brocade was discovered in Sussex – an immigrant arriving under its own wing power to initiate the process of colonisation. Thirty years ago, the only British location where you could reliably see it was Dungeness in Kent. Twenty years ago, it remained so rare and localised that a survey was commissioned of all known and potential breeding sites, from Kent to Hampshire. The largest colony, in Eastbourne, was later bulldozed for a supermarket development. This century, the Brocade's fate has improved. Perhaps voyaging inland along the spartan land of railway embankments, the moth has thrived. Its larvae gorge on naturalised Purple Toadflax sprouting in many urban gardens. It is now intermittently resident north to Norfolk and west into south-east Wales.

I am fond of Toadflax Brocade. Living in the capital during my inaugural summer of mothing, it was the first exciting moth I caught. Although I have seen three this year already – in Suffolk and Kent – I savour this London-garden example. Its Teddy-boy haircut reminds me of a Shark moth, another species I have seen on Dungeness shingle. Its thoracic crest recalls a Silver Y, our most famous immigrant moth. But it is smaller than either, and has silver and brown wings delicately patterned with white arrows and swirls, blazes and incisions.

Another egg-tray, another resonant moth. Roughly the size of a fingernail, Tree-lichen Beauty blends peppermint and grass-green with the blackness of coal and ribbons of white silk. On the foreign substrate of grey cardboard, it rivals a sore thumb for visibility. But as I ease it onto the trunk of an arthritic crab-apple tree, its colouration and pattern dissipates the moth across the bark.

Until 1991, Britain's experience of Tree-lichen Beauty comprised just three nineteenth-century records. It was a creature of the past, unattainable. During the 1990s, a dribble of immigrants became a gush. By the early 2000s, Tree-lichen Beauty was breeding across south-east

England. In 2006, it christened a trap placed on the roof of the House of Commons by Labour MP (and moth-er) Madeleine Moon. The Beauty's appearance was timely: three months earlier, Moon had tabled a parliamentary Early Day Motion on the parlous state of British moths, raising political awareness about an alarming new Butterfly Conservation report. Like Toadflax Brocade, Tree-lichen Beauty poses no environmental problem and provokes only enchantment. Their arrival and spread can be nothing but welcome.

We celebrate by popping a mile north to see our former neighbours. When we lived side-by-side in a Gough-Cooper terrace, the Alwan family might easily have been irritated by the blinding light I was running just feet from their patio on our handkerchief-garden. Instead, they were intrigued – particularly nine-year-old Matthew. Each time I caught a jaw-dropping moth – an Elephant Hawk-moth, say, or a Buff-tip – I knocked on the Alwans' door. Five years on, a now-towering Matthew and dad Ali run their own moth trap.

'I do moths perhaps once a week,' Matt says. 'I wish I could do it more. I just kinda like them. They're cool. While we were on holiday in Wiltshire last summer, it was my birthday. When I woke up, the first thing I did, even before opening presents, was check the moth trap.' It held a Privet Hawk-moth: 'the biggest and best moth I have ever seen,' he smiles. 'It was a wonderful birthday present.'

Estelle, Matt's mum, admits they're now all converts. She reckons the hobby appeals 'to Matt's desire to characterise, classify and organise'. Among other things, she welcomes moths as a cure for that quintessential teenage malady: sleeping in. 'Normally Matt wakes really slowly,' she says. 'But on a moth-trap day, he's up bright and early, heading straight out into the garden.'

Today, the tables are reversed: young Skywalker shows old Yoda a thing or two. Together, we sift through Pale Mottled Willows and Large Yellow Underwings, Matt diligently identifying each species. His most intriguing catch covers less than one-hundredth the surface area of the gargantuan birthday hawk: the exquisite four-millimetre Horse-chestnut Leaf-miner, one of Europe's three most economically damaging moths. The green leaves of an otherwise proud Horse-chestnut tree in a neighbouring garden are saddened by innumerable rusty blotches – the result of munching ('mining') by the moth's caterpillars. Given that this very tree once gifted me 2,500 adults in a single trap, it is hard to conceive that the world knew nothing of the species' existence until thirty-five years ago, when the first was discovered in Macedonia.

As a genus new to Europe, the moth was assumed to have originated elsewhere. But subsequent work – combining CSI-like investigations of DNA with the discovery of moths inadvertently pressed in herbarium specimens taken in nineteenth-century Greece – has revealed it to be a clandestine Balkan native, like the tree on which it depends. By 1989 it had spread north-west across Europe to Vienna, perhaps hitching rides on vehicles. It reached Britain – indeed, south London – in 2002. It now resides over much of England and Wales, particularly in urban areas where non-native Horse-chestnut trees have been planted, as they have been since the seventeenth century.

Horse-chestnut is among a scary 42 per cent of European trees species now considered in danger of global extinction. Alongside deforestation, fire and infrastructural development in its native Balkan range, natural and cultivated Horse-chestnuts across Europe are under attack from leaf-miners. Their defoliation reduces the tree's seed weight, germination rates and seedling vigour. Although this causes no lasting damage, the trees temporarily become so unsightly that authorities in some countries go to the expense of felling and removing them. But, in Britain at least, should we really fret about the fate of a non-native tree to the extent of vilifying a pretty little moth? After all, it's our fault the critter lives among us.

Returning to my parents' house, I am struck by the sight of a different type of moth trap. I hadn't noticed before, but every room has one, its innocent-white plasticity secreted inconspicuously along skirting boards or on windowsills. Having failed to see any textile-munching pests all year, I am suddenly staring at dozens of the things. All are quite dead. Lured to a sticky demise are numerous Case-bearing Clothes Moth. Each doffs a gingery cap, its grey wings faintly spotted black. In their midst sprawls a single mundanely beige Common Clothes Moth. The duo's banal appearance does nothing to redeem their actions. Dislike of the pair is almost universal, even among moth-ers: Butterfly Conservation provides advice on how to deprive them of a home. Perhaps another irony is at play: humans domesticated silk moths to create clothes; now another moth is destroying them.

The following morning, to the screeching accompaniment of non-native Ring-necked Parakeets, the little actinic trap in the garden reveals a more equivocal creature. Oak Processionary Moth's presence here juxtaposes the natural with the artificial, while its future provokes contrary opinions. First recorded in Britain in 1983, the first twenty years' worth of observations all related to natural immigrants – pioneer males seeking unchartered lands. But being sedentary, wild females are effectively unrecorded in Britain, so there is little chance of breeding and colonisation as a result of wing-borne arrival.

In 2003, however, Cypress Oaks were imported to the Royal Botanic Gardens Kew in west London. Unbeknownst to the organisation's arboriculturalists, the trees were riddled with Oak Processionary Moth eggs. These hatched to form larval nests, then trains of caterpillars ('processions') and finally adults. Discovered in 2006 – prompting a *Telegraph* report that 'a moth that can kill humans is found breeding in Britain' – this adventive colony soon spread across London into neighbouring Surrey. This matters because, like Gypsy Moth, Oak Processionary Moth can weaken trees – in this case, oaks – by repeatedly defoliating them. Like Brown-tail, 'OPM' larvae can shed irritant hairs that potentially cause people respiratory problems and skin rashes. This double whammy is fast transforming the moth into a public enemy, prompting authorities to spray 'contaminated' trees with a biological pesticide, *Bacillus thuringiensis*. But are such expensive programmes excessive? Did the City of London Corporation get value for money in spending £100,000 on tackling the moth in 2018? What about the national eradication programme, which cost £4 million during 2013–15? Or is the government's current import ban on oaks exceeding eight centimetres diameter a more effective long-term approach? Even the House of Commons' Environmental Audit Committee has considered the matter as part of a wider inquiry into invasive species.

Butterfly Conservation argues that threats from the moth have been exaggerated, with few reports of negative impacts on oaks, and Public Health England admitting the danger to human health was low. While acknowledging that control may be required in particular circumstances, the charity also waves a warning flag about adversely impacting non-target moths and butterflies, including rarities such as the oak-loving Heart Moth.

Somewhere there is a balance to be struck, but we're not there yet. In the meantime, I treasure this moment with a male Oak Processionary Moth. I admire his chestnut, densely feathered antennae (I can't help regarding them as 'eyebrows'); his woolly coat of a thorax; his grizzled, banded greyness. I think of him as a sixty-something physics teacher in a public school, wrapping himself up before retirement.

Even my mother appreciates this moth's agreeable gentleness. But it is a Jersey Tiger that really lights up her eyes. She has seen this feline-striped stunner before, of course. How can any south London gardener not? Her home lies in the epicentre of this moth's population explosion – a 600 per cent expansion in its distribution in two decades. For the first century since it colonised Britain from continental Europe and the Channel Islands – which it surely did, despite contemporary scepticism suggesting sleight of entomological hand – Jersey Tiger remained resolutely Devonian. Since the 1990s, it has spread east to Hampshire and north to Wales. By 2009, Jersey Tigers were brightening up south London – whether by natural expansion, human help or waves of immigrant-fuelled colonisation. From pavement nettles and thistles to garden buddleia, Jersey Tiger now finds street food galore here.

Gently, I fingertip-stroke the creamy-yellow stripes scissoring the moth's black wings. Instantaneously, it flashes traffic-light-red hindwings in warning. By day, this would shock any would-be predator. But Jersey Tigers are active throughout night-time as well in daylight and aposematic colouration serves no purpose when largely sightless bats are airborne. So tiger moths have evolved to 'listen' out for a bat's sonar through ear-like structures on their thorax, then take evasive action. Barbastelle bats, at least, have learned to respond by going quiet, enabling them to sneak up on moths unheard. In this remarkable ongoing evolutionary struggle, some tiger moth species now play bats at their own game, issuing ultrasonic clicks that jam the mammals' radar.

I sense the Jersey Tiger's superpowers are softening my mother's disdain for moths. The more moths she sees, the more she melts. If YouGov were to interview her, she would certainly now agree that moths are 'interesting'. I could seek to persuade her that moths are also 'important', talking of a keenly anticipated future where we harness moths' newly discovered pestilent properties for our own benefit, such as deploying Indian Meal Moth larvae to digest plastic. But, in the short term, that would probably worry her about the fate of her rice and cereal. Helping people see the light is best done one step at a time.

17

Winged wanderers

Kent, Sussex, Dorset and Cornwall
August–October

'Do you want me to bring it down before I leave at eight?'

It's just gone 6 am on an early August morning, and I am packing up after another night's mothing on the RSPB reserve at Dungeness. It's unsociably early for my phone to beep, which makes my mind yo-yo between disaster and elation. The message and accompanying photograph are from Owen Leyshon, a Romney Marsh Countryside Partnership ranger, who asks – rhetorically – if I want to see the moth he has just caught in his garden in nearby Littlestone.

The image discloses a boldly patterned moth with the angular form characteristic of geometers – as if drawn by a child with a ruler and protractor. Successive bands of cream, mocha and carob undulate across its wings. Dark teardrops are splattered fore and aft. I've never seen one before but nevertheless know precisely what it is. Owen has caught a Tamarisk Peacock. It's very rare and it's not from round here – a winged wanderer from overseas.

An hour later, a small welcoming party has gathered at Dungeness Bird Observatory ahead of Owen's gift-bearing arrival. Owen confesses that the rarity did not give itself up easily. Escaping from his moth trap, it alighted under the garage gutter. Recovering it involved hanging off a stepladder. He is understandably chuffed that the gambit worked.

Over the day, more admirers visit, accompanied by appreciative moans and clicking shutters. And for good reason: since the first in 2004, fewer than twenty Tamarisk Peacock have graced our shores. Kent has attracted the lion's share. When local moth legend Barry Banson bumbles in for a cuppa, I hand him the pot. After a few seconds' scrutiny, he grins: 'I caught two of the first ten in Britain.'

Barry is entitled to smile. Moth migration is simultaneously a wonder and a mystery of the natural world. That such fragile creatures are able to fly such distances, often across open water, simply astounds. It follows that for many moth-ers, catching migrants – insects that, by rights, live in another country but have somehow found their way here, paying no heed to political frontiers set by humans – provides the most thrilling dimension to their hobby. Just as birdwatchers chart the seasons through the toing and froing of migrant birds – the first Swallow that rarely makes summer, autumn's inaugural nocturnal Redwing whispering overhead – so keen moth-ers sniff out conditions favourable to immigration.

Steve Nash interrogates meteorological charts to predict when moths might depart their home countries and where in Britain they might arrive – then generously provides his interpretation via social media. He particularly looks for plumes of warm air emanating from the Mediterranean or Africa. Their progress accelerated by the wind, nomads may arrive on a broad front and travel far inland before dropping groundwards to rest. Nevertheless, many make landfall immediately after crossing open sea. This makes the English coastline between Suffolk and Cornwall the focus of attention.

Migrant-mothing is the art of the possible. In optimum conditions, almost anything could turn up. If opening your garden moth-trap makes you feel like being a child on Christmas Day, running traps at a site famed for migrants is all those Christmases coming at once. Some moth-ers, such as Matthew Deans at Bawdsey in Suffolk, stolidly dedicate themselves to a single site, night in, night out. Others chase weather systems. Self-confessed migration-obsessive David Brown once dashed from Speyside to south-west England on the strength of predicted conditions. I may not be as flexible as him, but I resolve to chance my arm at several coastal sites with a superlative track record of attracting migrant moths.

Protruding into the Channel like an obstinate teenager refusing to conform, the Kent peninsula of Dungeness is ideally placed to receive winged waifs. The ostensibly barren shingle sweep also boasts a community of like-minded moth-ers orbiting around the fulcrum of Dungeness Bird Observatory, where moth migration has been studied for decades. So if the oddities don't reach my own traps, those rolls of other dice may conjure a double six.

Just as well. The traps that Will and I run on our visits fail to produce any noteworthy immigrants. I say 'Will and I' but that's not strictly true. One August night, we trap separately, a mile apart, each with our own wingman. Even wing*man* is not strictly true, for whereas I chew the fat with Jac Turner-Moss, the Observatory's assistant warden, Will is accompanied by Sarah Morrison, a London-based naturalist.

A fortnight earlier, Will and Sarah's eyes had met over moth-heavy egg-trays at Dorset's Portland Bird Observatory. Tonight, Sarah arrives in Kent as Will's new friend. A day later, she leaves as his girlfriend, their amour illuminated by actinic. This will prove no temporary tryst. Within six months, lives entwined, Will and Sarah will move in together. They will become moths to one another's flame. Migrant mothing truly is the art of the possible.

Will and Sarah's union may be the rarest encounter of that particular Dungeness night, but it is not the only one. Indeed, the Observatory's refrigerator is full of them. Most people regard a fridge as a device for keeping food fresh. At Dungeness – and in many a moth-er's kitchen – it also serves to keep moths calm. Enter the Observatory and you are immediately upon what locals call 'the fridge of joy'. Today its stellar occupants are a brace each of Orache and Pale Shoulder. Both are proper vagrants, and lookers too.

The colour of a vigorously mossy lawn, with a pink-hued shark tooth on each wing, Orache used to inhabit Britain's Fenland, but drainage extinguished such joy during the First World War. Now exclusively a migrant, barely 120 have subsequently reached Britain. Of just two records nationwide in 2018, one graced my Norwich garden – evidence that nomads can occur in the unlikeliest of locations.

At first glance, Pale Shoulder resembles a sturdy piebald bird poo deposited by a thrush after a lavish meal. Closer examination suggests such scatology to be uncharitable. Each wing is strung up with fairy lights, a neon-blue infinity symbol occupying pride of place amid a swathe of deep chestnut. Barely forty have been seen in Britain before these two. Its normal residence is parched Mediterranean terrain with scant vegetation; perhaps the Dungeness 'desert' proves no egregious alternative.

We savour the moment – sharing air with such exalted lepidopteran company. Then realise that the fridge harbours an even more exotic gift. Prior to 2019, *Ancylolomia tentaculella* (a gigantic grass-veneer) was known in Britain from just half a dozen records. An influx this summer triples that total. The one in the Dungeness fridge is the second we see in two days, the other being a few miles away, in Bernard 'Clearwing King' Boothroyd's garden trap. Typical grass-veneers are underwhelming beige micromoths, raggedy grass seeds that wisp limply at one's feet. Yet *tentaculella* rebuffs such disdain, principally on account of its size and petulance. It is a Hagrid among its ilk, bossing the pot in which it finds itself confined.

This is thrilling stuff. And yet my excitement is contaminated with disquiet. These aren't *our* moths; we didn't catch them ourselves. These are someone else's Christmas presents. Worse, they are effectively captive: inside a pot, inside a fridge, inside a building. Seeing these über-rare moths requires no greater skill than opening a fridge door. Unless they expire in their plastic prison before we get there – a realistic prospect given their migratory exertions – we cannot fail to see them.

'Thrill of the chase' it is not. My unease at admiring the Dungeness trio of moths derives precisely from the ease of doing so.

I am not alone. 'Fridge-ticking' is a widening fissure in the mothing world. People routinely pop exciting moths into pots so that their friends can see them before the insects are released the following night. I have benefited greatly from this practice, this year and previously, and one acquaintance claims it has become particularly commonplace since 'birdwatchers started taking over mothing'. 'Twitching' birds – travelling out of one's way, sometimes by hundreds of miles, to see a lost feathered waif – is standard fare for thousands of birdwatchers. Those who relish such things regard twitching moths as an efficient way to extend their list.

A growing band of moth-lovers, however, argue that this approach deprives the hobbyist of the excitement of discovering their own moths, and the understanding that comes with that. It may reduce the incentive to 'go mobile' with moth traps, suppressing the appetite to investigate new areas and thereby obtain valuable conservation data. Moreover, the naysayers insist, keeping moths in pots for an unnecessarily long time is cruel – even an insecure attempt at exerting dominion. A growing cohort refuse to have anything to do with the practice, and scorn those who partake. 'I could no more tick a moth in a pot in someone's fridge than an animal in a cage in a zoo,' says Martin Fowlie, a conservationist whose blog is entitled 'Of Moths and Men'. 'I have even had to avoid potted Death's-head Hawk-moths at nature reserves,' he grimaces, 'so as not to sully my yet-to-be-had first encounter with this moth of my dreams.'

As I return the Pale Shoulder, Orache and *tentaculella* to their respective plastic confines, I feel my own attitudes evolving. I cannot deny the thrill I get from seeing a new moth in a pot, but the kick I get from catching my own is in a different ballpark. I resolve to make two changes to this year's mothing. First, the moth's welfare will be unfettered. I will release moths at the first opportunity – usually the night after capture – rather than hang on to them an extra day in case a friend wishes to see. Second, I will no longer go out of my way to 'fridge-tick' unless the opportunity is utterly irresistible – the massive, multi-coloured gemstone that is an Oleander Hawk-moth, say.

Staying at Dungeness Bird Observatory is about much more than access to the fridge of joy, of course. The greater rapture comes from running our own traps – and sitting alongside Jac and warden David Walker as they go through the Observatory's catch, diligently recording totals of migrant species to extend decades' worth of data. Their number includes common migrants with enchanting monikers: Silver Y (whose name is stitched into its wings), Dark Sword-grass, Dewick's Plusia and Scarce Bordered Straw. Each has a travel narrative of its own, a perilous voyage to relate.

David and Jac's traps proffer the odd rarity too. I swoon at a Golden Twin-spot that clings to one raggedy egg-box. A relative of the Silver Y, this plusia is uncompromisingly shaped. Large eyes dominate a head topped with a crown and wreathed by an Elizabethan ruff. A drumlin of a thoracic tuft lends relief to the gentle curve of its back before swelling upwards into a Munro. The new-minted form is wrapped in iridescence, with sheeny bronze, tin and gold courting the eye. A moth to behold.

But there is one frequent migrant that I crave to catch more than any other. Convolvulus Hawk-moth is a mighty, palm-sized beast – the same heft as our garden Privet Hawk-moths. So extended is this sphinx's proboscis – reputedly three times its body length – that early entomologists thought it must be related to a unicorn. Sharp-winged and solid, it is a powerful migrant; a hundred or more reach Britain most years. Yet the handful I have hitherto clapped eyes upon have all come courtesy of other moth-ers, extracted from pots and fridges, plus once, bizarrely, flying around a light at Wembley stadium during an England football match. Seemingly I lack what it takes to actually catch my own 'Connie'.

This summer, Twitter fills with images of garden moth-ers catching their first. Delight, joy and gratitude are widespread. Rachel Wisbey's tweet is typical: 'I just don't have the words to describe the emotion felt when I peered into the moth trap and saw this stunning hawk-moth,' she writes. 'So lucky and privileged I am. Thank you.'

There may be an odd reason for this Hawk-moth's affinity for gardens. It has an inexplicable fetish for clothing drying on washing lines. Accordingly, Owen Leyshon leaves his smalls out overnight in hope. 'The full washing load plus my underpants are going out on the line this evening,' he tweets. Sadly, it turns out Connie doesn't like Calvin Kleins. In September, celebrity gardener, author and burgeoning moth enthusiast Kate Bradbury has better luck. She is delighted to chance upon one clasping a cyan sock left out overnight: 'I had to stop myself from giving it a cuddle.'

At Dungeness later that month, I finally catch my own Convolvulus – and in circumstances as odd as Kate's. Returning to the Observatory in the early hours after running traps elsewhere, I am too wired to sleep. Instead, I slump on pebbly ground by one of the Observatory lights, sipping a decaf cuppa in an attempt to come down. Something skitters across the Observatory roof, then arrows vertically downwards towards the trap. Breaking its fall on my shoulder, it comes to a halt in my lap. I start, but it sits still. Defiantly parted wings reveal a fat thorax shockingly striped with pink, black and white, above which sits an alien face with glowing eyes outlined in Adam-Ant mascara. I have my Connie.

At least, I think I do. The following morning, when I eventually awake, David scorns my contention. He doesn't dispute the identification – just the assertion that it is *mine*. 'Whose trap was it attracted to?' he asks rhetorically. To count as my moth, David mock-grumbles, it needs to be in a trap that I have been running. Anything else constitutes theft. I take David's point. In mothdom, this is the way of things. I resume my search for an indisputable first-person Convolvulus Hawk-moth.

Bawdsey seems a good place to try. Sited on the north shore of Suffolk's River Deben and thus safeguarded from any expansionist intentions of the port town of Felixstowe, Bawdsey has become the east coast's prime site for Convolvulus Hawk-moth. Its pre-eminence is down to sixteen years of near-constant effort by Matthew

Deans. Following my flying visit to see June's Striped Hawk-moth (yes, in a pot…), 'Moth-hew' – as friends call him – invites me back.

I visit twice, each time running my own traps overnight near the estuary before meeting Matthew at dawn to go through the Hall's catch. This approach effects modest competition. In August, our respective migrant highlights tie for first place: we each attract a stunning Bedstraw Hawk-moth. These form part of an unprecedented influx, with hundreds being seen countrywide this summer. 'Never seen anything like it in all my mothing years,' Norfolk moth-er Neil Bowman tells me. Unlike the Norfolk coastal trio that the Barkhams and I encountered in June, the migrant status of the Bawdsey brace is unambiguous; the arrival is associated with the largest immigration of Painted Lady butterflies this decade. Both species most probably hail from Africa.

Will and I take time to revere both Bawdsey Bedstraws. As they rest, the experience is vastly different to those buzzy glimpses in Norfolk. The lightning strike bisecting the front upperwing demands initial attention. At the front of the broad stripe, a narrow white band blazes forward, continuing into an ivory eyebrow above a chocolate eye with a beady dark pinprick of a midpoint. An impression of alert watchfulness is enhanced by short antennae haring off at a slight angle, like resolute Deely-bobbers. The base of the hindwing is a riot of black, white and scarlet. The underwings echo the upperwing stripe, and the toffee-coloured abdomen is incised with neat, ice-white bars. They are splendid.

Our estuary traps elicit further intrigue. Two Oak Processionary Moths, the scourge of London councils, are migrants here. Matthew is excited; they are only the second record for Bawdsey. A Pine Hawk-moth slumbers in our saltmarsh trap, though there is no pine tree within miles. Where has it come from?

Over at Bawdsey Hall, Matthew's traps do better than he anticipates, given underwhelming conditions for migration and what he considers 'an overall poor year for migrants in Suffolk'. In the lee of an ivy-covered wall, beside a work shed where decaying armchairs slouch, a trap boasts a Scarce Bordered Straw. Moth fudge with a Zebra-striped hindwing, it normally occurs throughout the tropics, from South America to Australia.

Stumbling over Badger diggings as we criss-cross between traps, we keep adding to the tally of common migrants. For me, used only to the odd foreign visitor in my garden, the totals are impressive: a baker's dozen of Silver Y and forty Diamond-back.

I have carried a torch for both species since two prodigious events in summer 2016 that revolutionised my appreciation of moths. Mowing the garden lawn one June evening, I kept flushing up tiny moths. I realised they were Diamond-backs – effervescent critters barely half a centimetre long with a jagged silvery stripe where a vertebrate's spine would be. I had only seen a handful before, but scores now littered my lawn. The following day, on the Norfolk coast, clouds of Diamond-backs billowed up at my tread. Similar tales poured in from across Britain, from Cornwall to Orkney. Near Leominster, one person reported a two-mile stream of Diamond-backs: 'like driving through rain.' Thought to have left the Baltic on north-easterly winds, an unprecedented number had reached Britain. Several million were estimated to have arrived in Norfolk alone. As the Diamond-back flies, it is 1,000 miles from the Estonian capital of Tallinn to my Norwich home. Each moth would have had to fly a distance 230 million times its body length. Their journey is the human equivalent of walking from Land's End to John O'Groats and back 9,000 times without pause. 'Epic' is not word enough to encompass their migratory feat.

A month later, 600 million TV viewers watched Portugal's 1-0 extra-time win over host France in the Euro 2016 football final. The sport was dull but the spectacle enlivened by multiple thousands of Silver Y moths gusting around Paris's Stade de France. Its lights had been left on the previous night, forming an allure too great for the front of migrants to resist. As Portuguese captain Cristiano Ronaldo collapsed on the grass following a collision that would terminate his final, a Silver Y tended to his impeccably arched eyebrow. The image quickly became the tournament's most iconic, and dozens of 'Ronaldo's Moth' spoof Twitter accounts materialised overnight.

The scale of both events beggared my personal belief all the more for learning that the life strategy of both species is predicated on periodic mass migrations to maximise population growth. The Diamond-back is a moth so geographically successful that its range stretches from the Arctic (where a friend has seen one sitting on snow) to Australia. It has even colonised Easter Island – nearly 2,000 miles off the Chilean mainland. Meanwhile, Butterfly Conservation's Richard Fox explained the Silver Y pitch invasion to Patrick Barkham at the *Guardian*: 'Individuals must migrate and choose the right winds to fly at different altitudes to maximise the efficiency of their migration. There's very clever stuff going on.'

Moth 'migration' comprises four distinct phenomena. The first is genuine vagrancy: utterly lost insects blown off course by strong winds. The second involves pioneers: moths unconsciously expanding their dominion by venturing into unchartered territory. Such intrepid movements may eventually lead to a species' colonisation – such as Tree-lichen Beauty in recent years – but are a high-risk venture. Fail to find a mate and you've squandered the chance of succession. The third scenario is a 'push' factor imposed by inclement conditions – heat, drought and foodplant-failure, for example – that drive moths away from their usual home. The fourth is the purposeful strategy adopted by these Silver Ys: a multi-generational migration like that of the Monarch butterflies of Central and North America. Adult moths move north in spring, then stop to breed. Their caterpillars develop rapidly into adults then proceed north before stopping to breed. And so the relay race continues until the year's final generation, astonishingly, turns back south for the winter, minimising effort by harnessing northerly winds and flying at high altitude.

Little wonder that many moth-ers become intoxicated by migrant moths – some to the extent of using radar to track movements through montane passes such as Col de Bretolet in the Swiss Alps. Little wonder too that Mark Tunmore, editor of *Atropos* magazine and instigator of late September's National Moth Night, would choose migration as one of the two themes for the event's twentieth anniversary.

University of East Anglia fresher Jack Morris persuades me to celebrate National Moth Night by returning to Bawdsey. As in August, we double up. While running our own traps at the estuary, we also attend a public event run by Matthew Deans at Bawdsey Hall – one of thirty-odd opportunities countrywide. I have met Jack a few times previously – including on April's Belted Beauty survey in Lancashire – and am aware of his prodigious natural-history talent. Even so, his sharp eyes and deep knowledge take me aback. Jack spots things that I can't even see when he points them out. Moths that flummox me take him seconds to identify. I am quietly in awe.

That we are trapping at all is somewhat surprising. Our journey from Norwich takes three hours, double the expected duration, due to monsoon-level rains flooding every road going. But as we reach Bawdsey, the sky dries up and the wind is spent. As dusk envelops the Barn Owls supervising the car, our assumption of a no-go is replaced by quiet optimism. Perching on a trunk near our traps, a Delicate – a migrant as lovely as its name suggests – proffers hope.

Come the hour before dawn, however, that sole Delicate is as good as our estuarine traps get. Single Diamond-back and Rush Veneer are certainly migrants, and Angle Shades might be one. But that's it. Bawdsey might be the best east-coast site for Convolvulus Hawk-moth, but on National Moth Night we fail. We pack up and head to the Hall. Perhaps Matthew has had better luck. Perhaps Matthew has caught a Connie.

As usual, he has spent the night at home, comfortable in bed. His fellow National Moth Night participants, however, have kipped fitfully in their cars for a few hours outside the Hall. 'I'm too old for all this,' admits one. In the monotone of the autumn dawn, we head to the trap that gifted me June's Striped Hawk-moth and August's Bedstraw. 'There'll be a Connie on the wall, for sure,' I jest.

And there is. We don't see it at first because it's pretending to be a light switch. Then Jack, inevitably, spots it. Its size sends shivers of excitement down my spine. Bawdsey has excelled for Convolvulus Hawk-moth this year. So much so that the BBC programme *The One Show* asked Matthew if they could 'borrow' his catches for a story they were shooting. He obliged, packaging the moths in a cool bag to keep them calm and handing them to a researcher at Ipswich railway station. Fame indeed.

Before we adjourn for coffee and bacon butties, we count numerous other migrants. From a trap near a fallen Madonna, which turns out to be a prostrate, peach-coloured statue of Marilyn Monroe, we extract a Clancy's Rustic, named after Kent moth-er Sean Clancy, who caught Britain's first. New for me this year are two Pearly Underwing and L-album Wainscot. Fifteen Delicate rather eclipse our single example, and the list is bolstered by Silver Y, Dark Sword-grass and two Scarce Bordered Straw. For a National Moth Night premised on recording migration, that's a success. Across the country as a whole, the Night elicits 5,000 moth records across an impressive 304 sites.

There's just one problem, of course. Bawdsey's Convolvulus Hawk-moth flanks Matthew's trap, not mine. Judging from his book *Collecting: An Unruly Passion*, psychologist Werner Muensterberger might argue that my obsession with procuring my own is ultimately about 'warding off undercurrents of doubt and a dread of emptiness, even depression'. Nevertheless, I cannot stop. My idiosyncratic, self-obsessed quest must go on.

'You want to catch Convolvulus Hawk-moth? Try Durlston Country Park.'

I seek and receive advice via social media. Accordingly, in early October, Will and I head to Dorset's Isle of Purbeck. One of Britain's most famous mothing sites, Durlston boasts historical records back to 1883. Four weeks previously, one lucky soul caught twelve Connies here in a single night. How can we fail? With consummate ease, it transpires.

Arriving ninety minutes shy of midnight, we set up on craggy ground in the lee of a refurbished Victorian castle that serves as the Country Park's new visitor centre. Three traps have an aspect towards open sea, where Traveller's Joy filigrees the scrub. Two traps target an extensive tract of Holm Oak, a gnarled Mediterranean tree that has been introduced to Britain since the 1500s and has taken a liking to the east-facing side of the Park. We choose one location in homage to Bernard Skinner – the late, legendary moth-er credited for impassioning a whole new generation about moths – placing traps on the exact same spot that he did in 2012, below the steepling protection of a rock wall. If it was good enough for Skinner, it's good enough for us.

The evening starts chilly, with an oppressive easterly wind. But by midnight the breeze has abated. It is

Dorset's first really still night for ages. Our first trap round offers hope: Clancy's Rustic, L-album Wainscots and Delicates are already in. It's a night for sitting out with the traps, but – with our body clocks having shifted back to normal following mid-summer's nocturnality – we are simply too shattered to stay with the traps beyond half-one. To the tune of Tawny Owls, I collapse into the car and Will slumps into a sleeping bag on the dogshit-encrusted car-park verge.

A few hours later, Durwyn Liley draws into the car park while the sky is still thickly black. He wakes us with peaty coffee then we rush to extinguish our lights before the dawn dog-walkers arrive. There are signs that migrant moths arrived during the night, among them Durwyn's first Scarce Bordered Straws. There are plenty of Delicates plus a supporting cast of Silver Y, Rusty-dot Pearl and Pearly Underwing. Most intriguingly, we catch three Red Admirals – a butterfly that migrates at night as well as by day.

The best moth is a species that was unknown in Britain until 2008 but has since found coastal Dorset to its liking and has colonised. Given that its larvae feed on Holm Oak, Durlston was a savvy place for Sombre Brocade to ingress – a home from Holm, perhaps. We catch two, including one in the precise spot that Skinner used, which pleases us no end. As does the moth's appearance. Judging from its name, I was expecting the year's dullest moth. Instead, it is a confectionary box of chocolate. I count a dozen shades of cocoa on its wings, in a riot of shapes. The more I look, the more tones and forms I discern. A connoisseur's moth, this not-so Sombre Brocade.

But there is no Convolvulus Hawk-moth. I am succeeding at failure.

After packing up, we head west to the Isle of Portland. Targeting Connie here is no shot in the dark: on average, around twenty are caught annually in Portland Bird Observatory traps. Aware of our impending arrival, warden Martin Cade has put aside the week's star capture, a *Herpetogramma licarsisalis* (Grass Webworm). This Afrotropical micromoth has recently established itself in Mediterranean Europe and first reached the UK in 1998. In northern Spain, the population reached plague proportions in late 2018, with caterpillars defoliating vast areas of grass. Following such a demographic boom, and winds departing Iberia, it is not overly surprising that the past few weeks have seen fifteen records in south-west England, almost doubling the previous national total. Although rare, Grass Webworm is unremarkable in looks – so much so that it initially confused Martin: 'I was expecting something as large as a Mother of Pearl.' Instead, this mystery 'spread-winged shape' clinging to the underside of the Perspex cone was small – not much bigger than a Rusty-dot Pearl. Twice the moth dropped into the bottom of the trap only to eventually whizz to freedom, finding seclusion in a nearby ivy clump. Martin was more than a little relieved later that day when it re-entered the trap shortly after dusk. 'All's well that ends well,' he concludes with a grin.

To pass time we raid the Observatory fridge and sift through egg-trays in the old dustbin where Martin collates the previous night's catch for public inspection. The last time Will did this, he met the love his life. This time, we meet our first Cosmopolitan, a lissom, softly sand-toned wainscot. The moth's name derives not from cocktails but a distribution that spans the African, Asian and Indo-Australasian tropics. This migrant sometimes appears in mighty numbers; 500 were recorded in 2006. In other years – of which this seems to be one – barely a handful are seen. Three other moths catch our attention. There are hundreds of migrant Rusty-dot Pearls – 'routine numbers here,' Martin says, but more than I have ever contemplated. The other two are attractive Portland specialities offering intriguingly similar stories about our changing world.

Flame Brocade was resident in Sussex until the early twentieth century, before succumbing to national extinction. An occasional immigrant until 1990, it has since become a regular autumn migrant to the south coast. Some pioneers managed to breed, such that the species has become established back in Sussex – and at Portland, where up to seventy have been caught in a night. It is large and boldly patterned, with a silvery Santa's sock lying at right angles to an argentine blade that slices forward towards a blazing mohawk. All this on a shimmer of bronze and undercoat of mauve.

Will is even more chuffed by seeing several Radford's Flame Shoulder. Slimmer than the Flame Shoulder of our summer gardens, with violet tones and a vibrant flash running along the outer wing, it is a striking moth. It is officially a scarce migrant, exclusively visiting the English south coast. But numbers appear to be rising, with upwards of sixty seen annually since 2016. Martin, at least, suspects that Radford's Flame Shoulder has 'done a Flame Brocade' and is also now breeding locally. If so, intrepid migration can indeed pay dividends.

Late afternoon sees the skies open as a storm rages from the west. Less suitable conditions for mothing you cannot imagine – yet the air is mild enough for moth-migration guru Steve Nash to make the outrageous suggestion that the outlandish Oleander Hawk-moth might be on the cards. Nevertheless, we sleep through a windy dusk and the first exceedingly wet hours of a night. Only when the air calms do we wake and rush out to set up traps by torchlight. We choose a sheltered cliff and quarry on the eastern side of the island, treading carefully as the rain has gullied the path into a temporary river.

Every few metres, we spot moths hanging solitarily from grass stems, hoping to dry out. L-album Wainscots are outnumbered only by Lunar Underwings. Feathered Brindle competes on camouflage with Feathered Ranunculus; both species evaporate when popped on a lichen-encrusted rock. Multiple Angle Shades slurp rich juice at the blackberry bar. Enticed out by the rain, snails crunch sadly under foot. I wonder whether the snail can ever recover, can ever reconstitute its shell, or whether my clumsiness brings its demise. The wind cares not, continuing to roar overhead. We depart for bed.

Scant hours later, a chastisement of awakening ravens greets our bluster of a dawn. The gale continues to harry great rectangular breezeblocks of limestone that have long since slumped towards the waves below the quarried cliff. Despite the wind, our traps contain precisely 300 moths – a pleasing total given the conditions. There are ample Beautiful Gothic, an invertebrate stained-glass window that is a speciality of south-west England's grassy cliffs. We are delighted to catch our own Flame Brocade. Vagabonds are few but include – glory of glories – what is finally and indisputably our very own Convolvulus Hawk-moth.

I punch the air. Catching this beast is every bit as scintillating as I hoped. I lose myself in the blackwater pools of its eyes, take in every stripe, line and blemish, and follow the jagged leaps and falls of markings from the inner wing to its outer edge. I am complete.

As we dismantle the traps and lug them upslope to the car, I become bone-weary. Exhausted by the elation of finally trapping my target, I need sleep like never before.

But I don't get it. As I regain mobile-phone reception back at the road, I notice a text message from Chris Fox, a Dorset moth-er whom we met two evenings previously. The tone is matter-of-fact, the content unforeseen: 'I have caught an Oleander Hawk-moth. Would you like to see it?'

The world stops breathing.

Back at Dungeness in August, I resolved to resist pot-twitching wherever possible – 'unless it's an Oleander Hawk-moth or similar'. As an exception to the rule, it was a propitious moth to specify. We do the sums: returning home via Chris Fox's farm in north-west Dorset would add two hours to our journey. This seems an acceptable diversion for such a mighty creature. And more than worth forgoing the sleep my body is craving.

Come late morning, we pull into the village of Leigh. Chris welcomes us into a lovingly restored farmhouse that has been in the family for five generations. 'I didn't expect to see you again so soon,' he says. 'But given you were in Dorset…' Chris has the gentle yet authoritative air of the prep-schoolmaster he once was. We follow him along a wood-panelled corridor into an airy kitchen that reveres the light and he gestures towards an upside-down, lidless ice-cream tub plonked on a wooden plank. However temporary it may be, the two-litre residence is somewhat ignoble for such tropical royalty.

I remove the tub. And gasp.

Oleander Hawk-moth is peerless. It is a master migrant, having probably travelled well over 900 miles to reach Chris's farm; its core range is tropical Africa and Asia. It is huge – rivalling my mobile phone for size – and placid, even plasticine. It is a rarity indeed, with many of the 120-odd British records stemming from the first half of the twentieth century.

Irrepressibly patterned, its lysergic swirls giddy me. My eyes career forwards to the creature's head before turning towards the outer edge of the upperwing. At the very front of this, a dark-pupiled pseudo-eye stares angrily at me. Combined with an apparent pursed snout on the moth's thorax, it creates an uncanny impression of a hungry, floppy-eared and decidedly disgruntled goat. If I were a predator, I would be long gone. Even in monochrome, this moth would startle. But the colours render this moth otherworldly. A violet bruise pours into olive. Rose twists into apple-green. The antennae are lemon, those evil would-be eyes orange.

Yes, those eyes. There is an insidious malevolence in this beast's beauty. It is a not-so-jolly green giant. A witch doctor in invertebrate form. A bad trip. But it is addictive too. I cannot wrench my eyes away.

Eventually Chris breaks the spell, leading us away to the scene of the crime. As we pass a converted stable that houses his cider press, sheep regard us with an ungulate absence of curiosity. We cross a field to where Chris placed his moth trap. The Oleander was too large to enter, instead resting on a metal support structure designed to prevent the trap becoming a flying saucer in the previous night's gale.

We are the first to pay our respects to the best moth that Chris will ever catch. But we are far from the last. The Foxes make a lot of coffee that day.

How am I to follow the ultimate moth? Simple: try to catch one of my own.

A fortnight later, Maya and I base ourselves in west Cornwall for a full week of mothing masquerading as Dada-and-daughter time. We are joined by my sister and her husband, who appear one morning dressed as moths in recognition of my quest. In October, there is nowhere better than England's southwesternmost county for migrant-smitten moth-ers to base themselves. Given mild nights and Iberian winds, literally anything could turn up. My dreams are unfettered by reality: a Crimson Speckled for sure, perhaps a Silver-striped Hawk-moth. Why not a new moth for the UK? The time and place are perfect. But by the time we reach our rental cottage in the secluded hamlet at Cot Valley, a country mile from the village of St Just, all bets are off. The nights are clear and icy; a business-like wind hails from the north-east. These are not conditions to coax moths from the Mediterranean, nor weather to entice anything airborne.

I still try, of course. I run traps in the garden beside a crystalline stream. I secrete mobile traps about the cliffs, where they huddle below escarpments grouted with grass that are separated from the New World by nothing but the Atlantic. I spend cold, black hours ferreting among clumped, flowering ivy – autumn's most generous nectar source. All to no avail.

Perhaps I am unduly negative. After all, I manage to rustle up some acceptable nomads. Brawny and angular, a White Speck is new for me, a Cosmopolitan notable. Two Delicate and a Pearly Underwing are outnumbered by a dozen Dark Sword-grass. Among resident moths, I am taken with the rose-tinted cappuccino of Autumnal Rustics and confused by a Red Chestnut that has emerged six months before its normal flight season. But of proper rarities, there are none. Migrant mothing is the art of the possible – but so often the 'possible' remains precisely that.

18
Perfect blue

Suffolk, Kent and Hampshire
August–October

'I verily believe that the sight of a Clifden Nonpareil flying round my bedroom would make me leap from my deathbed.' Across all of P.B.M. Allan's wit- and scholarship-drenched writings on moths and moth-ers, his obsession with Clifden Nonpareil was clearest in the book *A Moth-Hunter's Gossip*. Sadly, no Clifden Nonpareil visited P.B.M. on his deathbed nor, to the best of my knowledge, did he ever clasp mortal eyes on a living *fraxini*, as he called it. Were P.B.M. to claim this immense moth that he describes as the 'greatest prize' available to moth-hunters, he foresaw being so overwhelmed with tension that he would fail to seize the moment: He expected to bungle the opportunity to catch the moth before it whisked away.

Despite Allan's comic exaggeration, many moth-ers will recognise the underlying sentiment. No British moth sets the pulse racing like Clifden Nonpareil. None possesses a greater expanse of wing. It is cryptically patterned with an unexpected, clandestine stripe of cyan that prompts its alternative moniker of Blue Underwing – a tone unique among British moths. Formerly extinct, it is now merely scintillatingly rare and prone to unexpected guest appearances. It is a moth that teases, torments and – for the lucky few – elates. As its name suggests, it is justly without equal.

The first Clifden Nonpareil I saw, incongruously, was flapping around Marcus Lawson's kitchen in Poole. Each downstroke was as muscularly precise and Jay-blue as a morpho butterfly from the South American rainforest. The second, somehow squeezed into a Tupperware container, was revealed to me by Keith Kerr in the margins of a birdwatching trip. Each encounter made me swoon. But these were twitches – orchestrated, guaranteed meetings devoid of the highs of discovery. I craved my own encounter with *fraxini* without an intermediary: just it and me.

For nearly two centuries, Clifden Nonpareil was but a sporadic vagrant to Britain. So desired were they as collection pieces that, in 1869, famed lepidopterist Edward Newman wrote in *The Natural History of British Butterflies and Moths* that Clifdens were regularly subject to fraudulent trade: 'Some supposed English specimens are sold by dealers at a very high price, a fact that holds out a perpetual premium to fraud.' Such underhand commerce, Newman suggested, made it 'next to impossible to unravel the history of every reputed British specimen'.

During the 1930s, Clifdens started breeding in Kent and probably in Norfolk. Perhaps P.B.M.'s plea – 'O Jupiter, grant that some day this lovely moth will adopt our island as a colony for his kind' – had been heard? The Kent cohort, in the famous Orlestone Forest, attracted much attention from moth-ers during the decade after the Second World War. One such visitor was Edgar Hare, who would journey down from London in the sidecar of a motorcycle chauffeured by his gardener. By 1964, however, Clifden was extinct – the victim of changing forestry practices. It remained a vagrant for the remainder of the century.

Then everything changed. Records surged during the 2000s, with localised recolonisation from Kent to Dorset. By 2018, a minimum of ninety-five populations were spread across eight English counties. Perhaps helped by a warming climate and widespread planting of *Populus* trees (poplars and Aspen, its larval foodplants), Clifden was back. Mark Tunmore sagely nominated it as a target species for the year's National Moth Night: the literal poster boy for the twentieth such event. Would I get in on the act?

I have an early opportunity to catch this perfect blue. In late August, James Hunter invites me to join him on a Clifden mission in Orlestone. Following family conflicts, I decline. Mistake. At half-eleven that night, one flumps into James's actinic trap.

As luck would have it, I do clap eyes on a Clifden Nonpareil that day. But it isn't *mine*. Worse, rather than amid a hauntingly ancient forest, it is in the shadowy car park of a scaffolded pub in a mid-Suffolk market town. David Hatton, a Hertfordshire-based blast from my student past, messaged me that afternoon from his holiday cottage in east Suffolk. Stumbling into the garden just after 6 am, David's bleary eyes clocked a humungous moth clinging to the trap's power cable. 'It was larger than the plug,' he recalled. 'Too big to even enter the trap.'

Being a newcomer to the world of moths, David didn't immediately recognise it. Furthermore, it was too large to fit inside any pot he had brought with him. So he took photos, let it be and returned to the cottage to flick through a field guide. After a couple of false starts, David twigged the moth's identity as a Nonpareil – 'a moth so big that it has a half-page to itself in the guide,' he said. Palpitating, David also realised his schoolboy error in not capturing this bolt from the blue, so rushed outside with a glass cooking bowl big enough to house the moth. Fortunately, it remained in situ and – perhaps to ensure David didn't make the same mistake twice – had half-opened his wings, divulging its unique Manchester City-blue stripe. It was indeed a Clifden. *His* Clifden.

And now, indirectly, mine too. As Beccles' thirstiest slip past us in search of evening ale, David discreetly opens a large cool-bag as if offering back-of-a-lorry goods. Swaddled in a blue-striped tea towel is a Pyrex bowl placed face down on a white chopping board, upon which – posing benignly, massively – is a moth bigger than my palm.

The sight jolts me. It is two years since my previous Clifdens. Two years in which I have forgotten how vast this moth is. And how striking. Even with the wings closed, its cyan-and-black baselayer modestly concealed, even with its ivory-and-jet underside hidden from view, the Clifden Nonpareil is gloriously patterned. There are silver and gold teardrops; stock-market volatility in tan, chestnut and lead; swathes of pale and bands of dark. All on a broad-based isosceles triangle that refuses to conform by allowing its two equal-length sides to gently curve. If I manage to ignore what looks to be a zombie Bart Simpson on the Clifden's thorax and inner wing, this is truly a mega-moth.

But there is a problem. Seeing David's Clifden is precisely that; it is *David's* Clifden. Scratching the itch is never going to be enough. As with Convolvulus Hawk-moth, I yearn to catch my own. I *need* P.B.M. Allan's 'greatest prize'. With the year throwing up numerous records of Clifdens in new and odd places – every moth-er and their mother seems to catch one – my chances have surely never been higher.

Within a fortnight, Will and I dedicate two nights to the Nonpareil in the hometown of British Clifdens, Orlestone. Nineteen were caught here in 2015, Bernard Boothroyd informs us. The previous year, his nearby garden was graced by eight. We obtain permission from Kent Wildlife Trust to trap at Smallman's Wood reserve – a privilege indeed, for Smallman's is arguably the best of the best at Orlestone. This is where James Hunter caught his Clifden. We have the co-ordinates of the tree beneath which he caught it. We will surely succeed.

We pull up at an innocuous five-bar gate where roadside trees are molten with enveloping autumn. The open understorey grants inspection of oaks and Hazel, birch and Beech, ample Aspen, trees I don't recognise. Thickets of Rosebay Willowherb stitch the paths. Dusk feels mild. Moths are flying. A Setaceous Hebrew Character hurries by. A Square-spot Rustic slurps on the sticky-sweet effluent of an overripe blackberry. We secrete traps as offerings at the toes of luminous Aspen, each slender trunk straining upwards to a froth of leaf. We drape Adnams-infused ropes over sober branches. We wait. We hope.

All to no avail. Our traps gently welcome the pleasant but not the craved. Gently arched Oak Hook-tips.

Numerous Oak Lutestrings – a new moth for me. Angle Shades by the half-dozen. The tiny, delightful *Metalampra italica*, all chestnut and cream – and a mega-rarity up until just a few years ago. But no Blue Underwing.

Then the night chills. Moths still; the vacated airspace becomes somniferous. No Clifdens are airborne tonight.

The second night is a carbon copy of the first: early promise, quickening cold, empty traps. The only blue is our mood. We could try a third night, but fear Groundhog Day. We cut our losses and head for home. En route, we learn that Clifdens were indeed on the wing in Kent last night. Three were seen elsewhere in Orlestone, while Barry Banson caught one near Dungeness. Now it is the air that turns blue with our foul-mouthed frustration.

David Hatton's catch in late August turns out to be part of a staggering series of records in the vicinity of Minsmere RSPB reserve (and, indeed, as we later learn, the best year in Britain for decades, if not ever). Twenty-something volunteer Henry Page even traps three on the final morning of his six-month internship at the reserve. 'Aspen must be aiding their colonisation,' he says of the farewell gift.

Henry introduces me to reserve warden Robin Harvey, who generously lets us have a crack at Minsmere's Clifden. But there is a catch: an overload of Hornets has placed the reserve's normal trapping site out of bounds. Thinking furiously, Robin suggests we try a sector of the reserve where moths have never been surveyed. He has been assured that it harbours Aspen, so we have a chance of Clifden. And even if we don't catch one, we'll be providing new information on what moths live in that sector of the reserve.

Ben Lewis, Justin Farthing and I take the deal on a Saturday night in late September. The omens are good: the night is forecast to remain a balmy 16°C, and that morning I received the latest issue of *Atropos* magazine, which stars a feature by Sean Clancy on the target for National Moth Night, scheduled for a week later. By half-seven, our traps are set up in a loose stand of Aspen at the segue between birch-dominated woodland and end-of-season heathland.

Two Hobby – parent and youngster falcons – scythe the sky in pursuit of tardy dragonflies. Below them, a Stone Curlew heralds the night with its otherworldly cry. I think back to when I last heard this bird's voice, at Weeting Heath in March while rummaging for Lunar Yellow Underwing caterpillars. Tonight, though I don't yet know it, I will catch my first examples of the adult moth. Back inside the forest, the air between the trees is collapsing in on itself. The sky above coagulates with the flow of Rooks and Jackdaws to their treetop dormitory. Justin departs for a family meal, promising to return later. Ben and I switch on the lights. Our time is now.

Just shy of 10 pm, Ben abruptly breaks off from our chat, his eye caught by a pale leaf drifting upwards amid the fulsome aspen canopy. Leaves have no powers of flight yet this one is contravening gravity. I catch sight of the 'leaf'; it is so huge and pale that I momentarily think it a squirrel. It then emerges into the open and confirms our jaw-dropping realisation that this is our Clifden Nonpareil. *Ours.*

For three scintillating minutes, we watch the beast mastering the forest through swoops and glides. It is a beauty parade. Admired from below, the Clifden's shocking silky-whiteness is besmirched by bold black bands. Briefly, it alights on an Aspen trunk – within feet of our bulbs and wine-ropes, but not close enough. It pauses with wings apart, a cleft of cyan visible. Then it is off again, airborne, spiralling upwards, disdainful of our ground-level temptations, ghosting back into the realm of myth.

'Hi guys! Anything yet?' Fed and watered, Justin has returned. Thirty seconds too late. He sees the elation on our faces but also reads the concern. He realises. His face remains stoic but disappointment saturates his eyes. 'Surely it will come back?' he asks hopefully.

It doesn't. Not by 2 am, when we grab some shut-eye in the Mazda hotel. Nor by 6 am when, aroused by a Woodlark's lullaby, we sift through all eight traps. They contain much to savour, but no Clifden to enthral. Using binoculars, we scan the length of every single Aspen trunk, looking for a roosting Blue. Several times, we lurch to a halt at a triangular knot underneath a protruding branch. In shape and pattern, this splash of trunk is such a dead ringer for a Nonpareil that evolutionary adaptation seems at play. But we convert none into a real-life moth for Justin.

Ben and I – at least – have found our own Clifden. And yet we have not. Because we did not catch the mega-moth, we remain contaminated by P.B.M. Allan's affliction. And now Justin is infected too. A Green Woodpecker laughs at our departure.

Social media continues to fill with ecstatic encounters with Clifden Nonpareils. The first for the county of Rutland, for Matthew Deans at Bawdsey and for the RSPB reserve at Lakenheath. Ten in Norfolk – the best showing since yesteryear. Simon Curson, a member of July's Speckled Footman team, sends me a photo of one he found nestling by a light on platform 3 at Brockenhurst railway station in the New Forest. Although superficially an astonishing stroke of fortune, I unearth rumours that this unseemly location might be a regular haunt for Clifdens lured in by overnight illumination. Really?

As chance has it, a fortnight later, Will and I are driving through Hampshire. Simon's railway station is only a few miles off route. Why not have a butcher's? Parking up, we mooch along the bridge that tiptoes over the tracks, then descend through autumn-slanting sun to platform 3. We have to start somewhere, after all. So it might as well be at the scene of the crime. At the bottom of the stairway, we turn right towards a strip light... and stop in our tracks. Plastered onto the brickwork within six inches of the illumination, just above head height, is a Clifden Nonpareil.

My brain blows a fuse. Words fail me. My knees too. Will staggers around, gasping. A Clifden. *Our* Clifden.

Remarkably, it is a different individual to Simon's – fresher, with blacker stitching along the trailing edge and gingery shoulder pads. The rumour is true. There is enough magic on Brockenhurst's platform 3 to rival platform 9¾ at Kings Cross. Clifden Nonpareil is justly a moth without equal.

19

The autumn garden – of memes and leaves

Norfolk
September–November

During autumn, a moth meme goes viral on social media. Its unidentified star is a chunky grey-brown creature, simultaneously ghoulish and cute, peering through a window, a crazed look emanating from glowing amber eyes. The caption – 'y'all got any lamps?' – makes a parallel between moths' predilection for light and an addict's craving for narcotics.

Spawning a series of Instagram gags, the meme is both humorous and apposite, reminding me of a joke by the late American comedian Bill Hicks about what moths bumped into before lightbulbs were invented. It also prompts articles – even on the *National Geographic* website – about the science of moths and light, providing welcome positive exposure for this insect underdog. 'Previously known as ugly, creepy, light-addicted butterflies,' writes Beckett Mufson on the webzine vice.com, 'the internet has all of a sudden decided it loves moths.'

From a British moth's perspective, the meme's timing is perfect. Come autumn, bar a few clingers-on, the airborne lives of our butterflies are spent. But adult moths power on, new species emerging every few days.

September proves the fourth-best month of our garden-moth year, with the average catch reaching two-thirds that of June's heady crop. On five nights across the first half-month, we catch more than 180 moths. But from September's mid-point, our total halves – the prelude to a dismal October where most nights fail to elicit double figures of species. November starts tolerably but plummets into nothingness: the only five evenings mild enough to justify using electricity produce just four moths apiece. But for its increasingly evident failings in terms of quantity, autumn's quality and character beguile.

As bewitched by light as the meme's protagonist, numerous Large Yellow Underwings riot around the garden. The blunderwings crash into walls, bulb and me, while those that end up inside the trap recall sugar-fuelled teenagers in a mosh pit. Maya is struck by the pea-green eyes of a smaller, more colourful relative, Lesser Broad-bordered Yellow Underwing. Their colour is a physiological adaptation to high light levels. This autumn, we learn that reflection-reducing protuberances in a moth's eye inspired the technology behind NASA's funky new camera capable of producing fine-grain images of astronomical objects.

Maya frets that our lamps overly disturb the moths. She worries we are being cruel and selfish. It is hard to deny that we disturb moths' lives. For all the meme's intrinsic truth, no moth actively desires to be deprived of opportunities to feed or fornicate. For a moth whose adult existence may last but a week or two, an evening's incarceration conceivably exerts a disproportionate impact. I offer Maya solace: unanticipated overnight accommodation spares moths the attention of predatory bats and, by discreetly releasing the inmates come morning, we avoid attracting birds' attention. Most moths we catch are wandering males; females are mainly unaffected, safely laying eggs elsewhere. We attract only a minute fraction of neighbourhood moths so our traps exert no population-level impact.

I sense this might be pushing it for a pre-breakfast chat, so change the subject with an overture of honey-topped toast.

As autumn unfurls, its elderly leaves curl, cueing the arrival of moths that look like leaves. A perennial favourite, Angle Shades, is back in force – the year's second generation of this army-surplus-store creature more numerous than the first. A quartet of gingery Thorns – August, Canary-shouldered, Dusky and Feathered – freeze with crinkle-cut wings raised aloft. My favourite is Feathered, ostensibly dotted with mould spores and wings hole-punched with age, the male displaying pennate antennae. I fear for all their futures; the abundance of each has at least halved since 1970. Dusky Thorn has fared worst, numbers having dropped by a seismic 97 per cent. Fortunately it seems to have turned the corner, its populations now picking up.

The most bizarre leaf-moth I encounter this autumn is discovered by Norwich moth-er Laura King, whose garden borders the River Wensum. Laura is new to the pastime but was nevertheless frustrated to fail to identify a fan-shaped moth, richly coloured chestnut, exhibiting an unfamiliar Morse-Code pattern of dots and dashes.

Baffled, she phoned a friend, Norfolk moth-er Dave Appleton. Confounded, Dave phoned another friend: Dave Norgate. Eventually the two Daves surmised that the winged mystery was a species of Australian lappet moth hailing from the genus *Anthela*. It had never been seen away from Down Under. How it reached Norwich is an even greater mystery than its identity, but an inadvertent stowaway amid a shipment is likely, given that no moth would fly 10,000 miles.

Nothing so special graces our garden this autumn, but nor do we lack excitement. A migrant Scarce Bordered Straw arrives the same night as the first of three Dewick's Plusias. The former is tatty after journeying hundreds of miles. Recalling a small, colourful Silver Y, the latter are ostensibly migrants too. But the freshness of their bronze shield and eye-catching, Tippex-white, mid-wing squiggle argues differently. It seems that Norwich has been colonised by Dewick's Plusia on the q.t., its larvae flourishing on abundant suburban nettle. It's another sign of changing times, witnessed without leaving home.

The Plusia may yet follow the trajectory of Blair's Shoulder-knot, a sliver of silver that brightens the odd October trap. It lies with forelimbs feeling forward, accentuating its already considerable length. The svelte moth was unknown in Britain until 1951, when Kenneth Blair caught one on the Isle of Wight. It became the third new British species to be named in Blair's honour. An insect luminary, Blair was a former president of the Royal Entomological Society and deputy keeper of entomology at the Natural History Museum. He died within a year of the shoulder-knot's discovery, eleven days shy of his seventieth birthday, and so didn't live to see the remarkable range expansion of 'his' moth, enabled by its larvae expanding dietary preferences to encompass Cypress cultivars beloved of British suburbia. Within half a century, Blair's Shoulder-knot had established itself north into south-west Scotland. It has since pushed even further, into Fife. No macromoth has colonised more of Britain in quicker time.

'Dada, why's this snake in our garden?'

It is hardly an expected question on a decadent, patio-sprawling afternoon, so I surge over to see. A scaly serpent the dimensions of my index finger is swaying from a willowy branch of a red-and-purple-flowered fuchsia. From its bulbous head swell two enormous black eyes, each with an expansive catchlight. Wendy Gray, a nurse, clasped eyes on the same beast on her patio in early September. 'It certainly woke me up after my night shift,' she explained to the *Hull Daily Mail*. The newspaper ran Wendy's story under the headline: 'Couple's confusion as snake-like creature with fangs and beak falls from the sky'. But this is no airborne snake, rather the sizeable caterpillar of an Elephant Hawk-moth.

The 'eyes' and swaying seek to disconcert predators. The bulging head is false, the caterpillar having retracted three frontal segments to expand those lying behind. When those same portions are extended, they resemble a trunk, explaining the moth's name. This beast is part-moth, part-serpent and part-pachyderm. It's the first time we have seen the larvae in our garden, despite catching innumerable adults. In this, I am not alone. For P.B.M. Allan in

A Moth-Hunter's Gossip, only the caterpillar's odd habit of crawling upwards 'for a breather at tea-time' enabled moth-ers to encounter him. It's a sure sign that summer is over when the next generation is out and about.

After this agreeable larval interlude, we revert to adult moths. The most swoonsome day of our garden-moth year falls on 14 October. This is the morning that the first Merveille du Jour freshens our trap. It is sturdy and peppermint, wings scrawled with black and etched in white. It seems overly showy, a sure-fire meal for a hungry bird. But pop the moth on its preferred domain, a lichen-encrusted surface, and it dissolves – camouflage complete. Such duality characterises this species. The adult flies in autumn yet its scientific name, *aprilina*, speaks of the loveliness of spring leaf burst, while its common name – meaning 'wonder of the day' – falsely suggests this concerted night-flyer to have diurnal habits. This is an intoxicating moth.

No less winsome, in its own way, is Black Rustic. This moth has texture: its blackness is stroke-me-now velvet – the perfect creature to fly at Halloween. But its darkness conceals contradictions: the moth lives across much more of Britain than in 1970, yet its abundance has plummeted by three-quarters. Surprisingly, this year proves bonkers for Black Rustic. 'Far more than normal,' many people comment, proud as punch. One night, a Norfolk moth-er catches 155. It is a record year everywhere… bar our garden, where we catch forty all-in.

Some nights Maya persuades me to hang out tacky ropes saturated with treacle and rum ('they smell so good, Dada!'). It pleases me no end that Old Lady moths get stuck into the boozy sugar. A Satellite orbits cautiously, while Lunar Underwings cram along the dripping lengths of cord. Better there than in the underwear drawer of author Amy-Jane Beer, where one seeks solace. 'It's a Lunar Underwing,' I reply when Amy quizzically sends me a photo of the moth. 'Clearly got its name slightly wrong. Either that or you have an underwing drawer.' Delighted, Amy promises to rename the place where she keeps her smalls.

Halloween past, autumn starts sniffing at winter when the trap becomes dominated by moths named after months. November Moths are disc-like greyness, all surface area and no substance. December Moths are the opposite: hulking, hairy and black. With its buff brow-band, seen front-on, the resemblance to a Musk Ox is uncanny. This stout, swarthy moth is the last garden species to which I put a name this year. Maya and I have run traps at home on 151 nights, recording 14,000 individuals of 466 species over the year. This is treble the diversity of all other animal species I have seen in the garden *put together*: five years' worth of birds, mammals, butterflies, dragonflies, hoverflies, beetles, bugs – the lot. And still our moth list is under half that of some moth-ers' gardens. There is much yet to discover – without even leaving home.

20
Southern comfort

Kent, Essex, Dorset and Suffolk
September–November

For my daughter, at least, no mug of hot chocolate is complete without marshmallows. Now made from egg-white beaten with gelatine and sugar, for centuries previously these squidgy, pastel-toned, tooth-achingly sweet cubes were concocted from the candied root of Marsh Mallow plants. In south-east England, this now-scarce plant grows almost entirely amid coastal marshes separating the seaside towns of Rye and Hythe. It was not until 1951 that entomologists exploring levels near Hailsham chanced upon a moth that shared our taste for Marsh Mallow, its caterpillars feeding exclusively on the plant's roots, stems and rhizome. The moth was new to Britain and otherwise occurred only very locally in a handful of European countries. Precisely, if unimaginatively, the discoverers named it Marsh Mallow Moth. It is one of five southern specialities that I target during the year's final third.

Being dependent on a single foodplant is not unusual; specialist niches fertilise the immense diversity and ecological success of moths. But dependency on a hen's-teeth foodplant makes for a precarious existence. Any serious shift in the plant's standing would likely have critical importance for the moth. By 1993, agricultural practices, dyke-widening schemes and perhaps a drier climate had taken their toll on the Marsh Mallow plants. In Sussex, few were thought to exist – and the two biggest stands were no larger than a typical kitchen. Its fate tied to that of the plant, the moth's fortunes plummeted. On Romney Marsh, the presumed stronghold, the moth wasn't seen between 1978 and 1993. Between 2001 and 2016, populations slumped by two-thirds. This endangered moth now survives only at a handful of small colonies that pimple East Sussex and Kent. It's a creature in need of help.

Nobody has invested more personal energy in understanding and conserving Marsh Mallow Moth than Sean Clancy, who first got into moths aged thirteen when he joined a school birdwatching trip to Yorkshire's Spurn Point in 1976. 'At the bird observatory,' he recalls, 'I first saw a moth trap in operation – and was immediately hooked'.

Sean cites two key reasons moths provide never-ending fun and intrigue. The first he describes as 'the randomness of moths', citing in evidence recently spotting a micromoth new to him in the prime habitat of the inside of his bathroom window. 'The second,' he says, 'is that anyone can catch a "mega-moth", so a total newcomer to the hobby can catch a species that someone with a lifetime's vocational exertions has never seen.' As someone who announced his arrival to the community of Norfolk moth-ers by stumbling into (and somehow failing to recognise) a Crimson Speckled, a rare and ostensibly unmistakable Mediterranean vagrant, I very much understand Sean's point.

Will, Jac Turner-Moss and I meet Sean shortly after dark in the second week of September. We gather at the entrance to a farm near Rye, on the western fringe of Romney Marsh – the only site where Sean can guarantee seeing Marsh Mallow Moth. We follow Sean's tail-lights deeper into the farm, bumping along a track that stands proud two metres above low-lying grassland.

Descending from the car, we swing torches into the blackness. Sean is already off; this is one of several sites he will survey tonight, so he has no time to lose. Dawdling behind, we spy some chest-high plants over to one side and push through tall, rank grass towards them. They are thrusting Marsh Mallow, each with a rigid, upright stem and fulsome, pointed leaves with a serrated edge. The five white petals are a thing of the past, succeeded by tightly packed ripe fruits.

Within three seconds of reaching the plants, Will's headtorch illuminates the namesake target. 'Here's one!' he calls, as a chunky pale moth weaves slowly between Marsh Mallow plants. Three seconds. If only every quest were so straightforward.

In the field guide, Marsh Mallow Moth looks mundane – a moth you feel you *ought* to see because of its scarcity, even though it looks as boring as hell. But field guides can be misleading. You shouldn't necessarily judge a moth by what you see inside a book's covers. Marsh Mallow Moth proves to be surprisingly fetching. The tone of a sand dune at dawn, threaded by golden veins, with latte contours, milk-chocolate kidney spots and a golden leonine mane. A subtle beauty underwritten by rarity, this moth delights.

Wandering between clumps of Marsh Mallow plants, we soon count upwards of twenty individuals. Males hurtle overhead, desperately seeking mates. Females fly less hurriedly, whirring clumsily like ghostly cockchafers. The night is cool so the moths are docile once landed. One alights on Will's beard. Another adorns Sean's eyebrow. I sport one as a brooch. We play voyeur to three copulating pairs, each conjoined while clinging to a flowerhead or spear of grass. Smaller than his mate and with thicker antennae, the male draws the short straw, facing groundwards and holding on, refusing to let either gravity or love tear them apart. Another female has already done her business and is delicately squeezing her swollen abdomen to eject eggs into a coarse rush. A further female is less successful: we discover her trapped in an orb-spider's web, already injected with venom. Sean's mighty challenge has got one moth harder.

As I observe a third female lay eggs on a Marsh Mallow stem, an irony strikes me: if the moth procreates super-successfully, its caterpillars will munch every bit of Marsh Mallow plant going. At which point there is nothing left for future generations to eat. The species eliminates its own habitat. Success breeds failure. Fortunately, Sean explains, the Marsh Mallow Moth seemingly has a trick up its pupal sleeve. When the larval density is unsustainably high, the adult moth disperses to pastures new. In other words, if a Marsh Mallow Moth caterpillar meets too many other caterpillars as it eats Marsh Mallow roots, it knows to go on a journey once it develops wings.

The risk is that the foodplant is so rare that there's little chance of an adult moth discovering a new, unpopulated tract. Accordingly, Sean and others are identifying suitable areas to translocate Marsh Mallow.

This seems to work. The previous week, Sean was elated to discover three Marsh Mallow Moths on painstakingly cultivated plants at Rye Harbour reserve in Sussex. At a marginal site in Kent, Sean has advised contractors to reprofile a bank to reduce its gradient, which should allow more plants to take root. 'The location has loads of potential,' he enthuses, 'so I hope we can get numbers back there.'

Sean has also worked out a system to help the moth at its strongholds: once the plants have died down, landowners introduce sheep to the Marsh Mallow. The herbivores graze competing vegetation – notably thistles – hard, hopefully without developing a taste for the roots of Marsh Mallow itself. I ask Sean how he convinces landowners to help.

'They get paid,' he says simply. For these farmers, moth-friendly practices are a condition of the Countryside Stewardship Scheme payments that the government has made under the European Union Common Agricultural Policy. This cues muttering among our ranks. What will happen to such payments after the UK has parted from the EU is unclear. Will we still be able to afford to financially support farmers? If so, will public money be used to buy public goods or squandered on butter mountains and vast estates?

The year 2021 marks the ruby anniversary of the Marsh Mallow Moth's discovery in Britain. With such a constrained life cycle, it will never be a common moth. And in a post-Brexit Britain, it faces an uncertain future. Two months before our visit, a BritainThinks poll found that the environment has become the public's joint-third priority issue, behind only health and poverty, and above terrorism, crime and jobs. But with even the most positive of models predicting an economic downturn in the short term at least, what price the air we breathe, the water we drink and the wildlife whose country we share?

One thing is sure, though: with conservation heroes such as Sean Clancy serving as their guardian and its inclusion in Butterfly Conservation's project 'Kent's Magnificent Moths', Marsh Mallow Moth has a fighting chance. And that might just be enough.

On the north coast of Kent and particularly along the North Sea fringes of Essex, conservationists are securing the future of another searingly rare moth. This one is so rare that any form of intentional disturbance – such as surveying through light-trapping – is illegal. This presented Kent moth-er Andy Taylor with a problem: the autumn after he moved into an immaculate 1960s period house in clifftop Tankerton – a residence so startling that it is regularly hired by commercial photoshoots – he started catching Fisher's Estuarine Moth in his garden. 'My first reaction was to fist-pump in jubilation,' he recalls. 'But then I felt guilty at disturbing a Red Data Book species.'

Each September or October evening when Andy flicked the switch on the two traps in his expansive rear garden, or on the two facing the sea on his elevated terrace, he risked breaking the law. Eventually, pressure from peers forced him to obtain a government licence allowing him to legitimately catch moths in his own garden – perhaps a unique situation. He typically notches up seventy Fisher's Estuarine Moths each autumn, and up to thirty-six in one notable crazy night.

Fisher's Estuarine Moth is not just very rare, not merely legally protected; it is also enigmatic and controversial, strikingly handsome and a conservation success story. The first British example was found on a kitchen window by Ben Fisher in 1968. Fisher happened to be volunteering at the Natural History Museum so he asked colleagues to identify it. They determined it to be a European species – and honoured Fisher with the moth's English name.

Like Marsh Mallow Moth, research showed that this 'newbie' appeared to be an overlooked native confined to rough grassland harbouring a single, rare foodplant with a restricted distribution: in this case, Hog's Fennel, an umbellifer. Remarkably, the moth was thought to rarely stray more than ten metres from the plant. If true, this would make natural population spread impossible. The species was trapped by its own ecology.

It thus came as a surprise when, during 2001, Fisher's Estuarine Moth was discovered on patches of Hog's Fennel on the north Kent coast. Opinions remain divided about how it came to be there. Andy favours natural colonisation – subsequent research has proven that the moth *can* disperse relatively widely. Others believe collectors nabbed larvae or eggs from Essex and illicitly released excess breeding stock onto Kent Hog's Fennel. Atonement for their theft, perhaps, or a genuine desire to help the moth's conservation prospects? Either way, the only British moth ever to have enjoyed European-level protection, through the Habitats Directive, now quietly thrives near the bohemian seaside resort of Whitstable. It makes a logical first call.

Will and I step out of Andy's house an hour after dusk. The night is cool, but still. Within seven seconds, we have reached Fisher's Estuarine Moth territory. We stand atop a thickly grassed slope that steeples fifty metres to the promenade. Beyond is the oily sea – flashing red lights of wind turbines far ahead, the tangerine blur of the Isle of Sheppey to our left. A Ringed Plover – the short-legged shorebird – whistles from the unseen beach. *Poo-eee* it calls, the second note higher, inviting hope.

We search the slope methodically. Top to bottom, section by section. Thickets of foodplant are scattered along several hundred metres of tussocky ground. I trip constantly in runnels, my eyes fixed on the tough, dry stems of Hog's Fennel rather than assuring my footfall. The massed, splayed lemon-lime florets of summer are long gone – parched into tawny antiquity. We search the slope rigorously for two salty hours but nose out no Fisher's Estuarine Moths.

We call it quits and return for coffee. Andy takes the opportunity to show us the one he prepared earlier. He caught his first Fisher's Estuarine Moth of the season the previous night and – to keep it safe from daytime predators – has looked after it until nightfall, before returning it to a Hog's Fennel stem. We have not found our moth, but we still get to see one.

It is alchemy in moth form. Wings of tin, but swathed in bronze, circled with silver and glittering with gold, this is a moth by Klimt. It is heftier than imagined – the size of a Dark Arches, a familiar summer moth. There is something of the Norse warrior about it, a creature before which to kneel.

Which makes it all the harder to understand the conservation world's apparent history of apathy towards Kent's Fisher's Estuarine Moths. Those moths flying on the Essex side of the Thames, Andy explains, have benefited from long-standing study, habitat-management protocols, foodplant translocation, diligent captive-breeding and painstaking reintroduction. Here in Kent, until very recently… nothing. For 'FEM', as the moth is nicknamed, it has long seemed that TOWIE. The ostensibly protected slopes at Tankerton receive only a modicum of habitat management. Worse, there is no restriction on public access: summer parties, fairs, even triathlons are common. A single prematurely discarded portable BBQ, and the Hog's Fennel and its moths would be no more, Andy fears. FEM deserves better.

Accordingly, it is most heartening that Fisher's Estuarine Moth forms part of Butterfly Conservation's Kent's Magnificent Moths project, alongside Marsh Mallow Moth, Black-veined Moth and other stars of this year. Conservation ideas on the table include habitat extension through planting cultivated Hog's Fennel, structured surveying and – as with Marsh Mallow Moth – provision of advice on Countryside Stewardship options. Finally, Kent's FEMs are being given the attention they deserve.

Tankerton could have been mission accomplished. Fisher's Estuarine Moth: been there, done that. But it doesn't feel enough. I want to see more – to understand this moth in its uncontroversial, conservation-infused Essex context. So five days later, I team up with Norwich moth-ers Justin Farthing and Dave Holman to join the annual count of Fisher's Estuarine Moth at Beaumont

Quay, on the edge of Hamford Water, barely a mile from the site of Ben Fisher's discovery.

As our baker's dozen of surveyors congregate, a red sky – a memory of the previous night's harvest moon – is partly obfuscated by straggling clouds. We hail from far and wide, but none more so than a Kiwi accompanying an English friend who lives in New Zealand. Also stepping out of the murk is Chris Balchin, whom I know from birdwatching in South America. It's news to me that Chris has been quietly into moths for forty years. As we exchange greetings by a deep muddy creek pockmarked by stranded dinghies, so do other creatures. Lapwings *pee-wit* and Curlews call their name to a backdrop of the scissoring and chirruping of an orchestra of crickets and grasshoppers.

We turn away from a run-down barn, tread past *Mrs Elizabeth Bennett*, a supine, rust-weeping boat, then galumph along half a mile of elevated sea-wall. An information board proudly announces the presence of our winged star, but its undercurrent is politely clear: please don't trash this place, please don't trample its plants. To our left, the land slips away into estuarine gloop, from which a Redshank bleats. To our right, rank, low-lying grassland is incised by ditches and hemmed by hedges; two Little Owls bicker.

We are convened by Zoë Ringwood, who has dedicated much of her life to saving Fisher's Estuarine Moth, initially through her day job, then in addition to it. She guides our survey: 'We'll spread out for thirty metres inland of the saltmarsh then track the sea-wall back. The moths perch on or near Hog's Fennel. Look for their red eyes glowing back at your torchlight. We don't need a licence to look for moths but it would be illegal to disturb them by getting too close. I don't know how many we'll see – if any – as we're varying things by surveying a week earlier than normal.'

The air is muggy, almost smoky. Those who prefer easy walking take the sea-wall, scanning the fringing vegetation. I wander through leek-green reeds onto a Couchgrass-swamped bank. Ahead of me, someone shouts: 'Found one!'

We swiftly assemble. 'It's a female laying eggs,' Zoë says gently. 'Give her some space.'

I just about discern the female moth's egg-laying device – an ovipositor – with which she probes between the grass stem and the outer leaf-sheath. Here she will deposit her shot at succession. The moth will probably have emerged from her pupa the previous dusk, Zoë explains. After pumping up her wings for three hours, she will have mated after midnight. Tonight she lays her eggs. Unable to feed, she will die within a week. Adults are pure sex machines; adult life is all about descendants.

Someone spots another moth. Zoë arrives to make the formal record, whispering into a dictaphone. 'Section three. Borrowdyke edge. Male resting on Hog's Fennel.'

A Redshank continues to call, unseen but nevertheless witnessed, from a distant mudbank. A Dunlin wheezes overhead. Both shorebirds are reminders of our broader location – easy to forget when our world is temporarily confined to the arc of a headtorch.

Location explains everything when it comes to Fisher's Estuarine Moth conservation. This insect lives on the edge. Its main population, on Skipper Island, was inundated during a tidal surge in December 2013. It is recovering, but far-sighted conservationists had already enacted contingency plans. Since 2006, Zoë and others have worked with farmers to develop a network of brand-new sites for Hog's Fennel across coastal Essex, establishing tens of thousands of individual nursery-grown plants. Colchester Zoo has captive-bred moths for release at these new sites. And the Environment Agency has adopted a mowing regime for the Hamford Water sea-wall that integrates the moth's interests alongside flood prevention.

I learn much of this from a man bringing up the rear of the group. This is Leon Woodrow of Tendring District Council, our official host for the evening. 'I've been doing this for twenty years,' he smiles. I ask him why. 'Zoë,' he says simply. 'I started by helping her on health and safety grounds. But I got caught up in the whole thing. It's become an integral part of my autumn.'

If Leon is hooked, then what is Zoë, the instigator of all things Fisher's? 'The moth has got a hold on me,' Zoë admits. 'As a researcher, it used to be my full-time job. While working for Natural England, I set up the programme helping farmers manage their land for the moth. Now it's just a bit here and there, to keep my hand in.'

And how Essex's star moth still needs her. After she has registered the night's tenth Fisher's Estuarine Moth, my friend Chris Balchin approaches her. He has been keeping an eye on a patch of Hog's Fennel near Colne Point, forty-five minutes' drive away. Chris spotted his first FEM there the previous night. Zoë encourages him to survey every week for the next month and allays his concerns about walking through the habitat. Chris is buoyed. Zoë has a convert – a new recruit to her platoon of Fisher's friends.

The discoverers of Britain's first Southern Chestnut – encountered on a Sussex common in 1990 – believed this predominantly western Mediterranean species to be a long-established resident. It had, they reckoned, simply been overlooked by dint of flying in a landscape rarely surveyed in autumn. A few years later it was found widely across the New Forest and on Dorset's heathlands too. These findings strengthened the alternative theory that Southern Chestnut was a recent colonist spreading its range. Whatever its origin, Southern Chestnut's apparent rarity ushered it into the Red Data Book of nationally threatened British wildlife.

For this noctuid moth, timing is everything. The window of opportunity to see it is as tight as anything we will attempt this year. Adults are on the wing for barely three weeks in mid-autumn, and their daily flight is usually over within an hour of nightfall.

'When we try for this moth,' Chris Thain of the Moths of Poole Harbour project told me a few weeks earlier, 'we start early and are usually packed up and down the pub by eight o'clock.' My path intersected that of Chris and his colleague Abby Gibbs during July when we searched Wareham Forest all night for Speckled Footman. Several of that night's gang reunite for tonight's search, keen to resume collective wilderness-mothing alongside individual garden endeavours. We meet at Slepe Heath, where I also ran traps in July. Ten weeks on, the Bell Heather and Cross-leaved Heath have almost forsaken their magenta vibrancy, their flowers shrivelled ginger, their lead-grey stems brittle, their preparation for winter almost complete.

There are enough of us for Abby to suggest splitting into two groups that survey different sides of the heath. She and four others remain at Slepe, while six of us hare round to the east flank of Hartland Moor. With the evening inking in rapidly, we have precious little time to cart our gear along the old, sunken tramway onto the heathland. By the time we have illuminated the surroundings and splayed gloopy wine ropes over gorse bushes, moths, including a migrant Vestal, are already flailing around.

It's only when we draw breath that I realise quite how cold it is. The peat-pool sky offers no safety net for ground heat. There are no clouds, few stars and barely a red slither of moon from which Jupiter and Saturn stretch southwards. Helping me run one set of traps, Phil Saunders shivers as a northerly wind crisps the air. Wingman Will jogs to the car, seeking hat and gloves. We plume dragon rings. These are not moth-flying temperatures, and the recent soggy weather will have done little to induce Southern Chestnuts to emerge from their pupae. Abby and Chris have yet to catch one this season. I am not optimistic.

'Down the pub by eight.' Chris's words reverberate around my mind. It's already past beer o'clock. We haven't seen a single moth for half an hour, and the traps are resolutely vacant. At half-eight, things perk up – albeit not for us. My phone pings with a message from

Abby: '1-0.' A terse celebration, imbued with implicit, literal one-upmanship. She's caught a Southern Chestnut! Even if – or seemingly *when* – our group fails, we will mosey from Hartland to Slepe and admire this nationally threatened moth. We are buoyed but also jealous. Why have we been shunned?

Without opening the traps, we continue to peer among the egg cartons – but now forlornly rather than in hope. There is nothing whatsoever in the first four traps. A couple of Horse-chestnuts slouch in the next two. Phil and I open the penultimate trap: ten egg-trays are unfettered by moth. We have but one left, again resorting to supping at the last-chance saloon. And again our endeavours are rewarded. A Southern Chestnut rests incuriously in the final carton. When this clandestine creature arrived is unclear; we neither noticed it fly in nor spotted it while perusing the traps earlier. But we whoop excitedly. We have our own. I tap the moth out into a pot... and miss, the supine insect sailing downwards towards the dark, impenetrable ground layer. Fortunately it lands on Phil's fingerless glove instead. We still have our own.

Southern Chestnut is understated but lovely. In the book it looks underwhelming: on the red side of conker but utterly unremarkable. In life, it has sharply pointed Titian wings, an enchanting scarlet face and chestnut legs. It is an unexpected looker. Delighted, we consign its image to our digital memory cards, then encourage it to crawl into the heather jungle.

We regroup with Abby and co., comparing notes. Our Southern Chestnut, Chris Thain says happily, is the first for Hartland Moor. 'It fills in a gap in this moth's distribution.' Abby and Chris have caught two at Slepe. All three are freshly emerged; a night or two earlier and we would probably have been unsuccessful. Narrow margins indeed. Had the night been warmer, Chris thinks we might have caught double figures. As it is, we did well.

While we relax with top-notch cider made by attendee Chris Fox, the Moths of Poole Harbour pair reflect on how their initiative is seeking to win hearts and minds for moths. Regularly running public events to challenge preconceptions, they find that many people have never really looked at a moth. Across two weekends at Corfe Castle – a 1,000-year-old fort that received 237,000 visitors in 2018 – Abby and Chris Thain introduced 3,000 people to live moths.

'First up, you have to tell people that the moths aren't dead, can't sting and don't bite,' Chris explains. 'Then you show them pretty moths – and watch them get excited. You actually only need two species. Any hawk-moth, but particularly a pink Elephant, which people think is some mad tropical creature. And a Buff-tip. People can't believe that it lives in Britain rather than the Amazon.'

Mention of this master of camouflage prompts Marcus Lawson to recall running a moth trap at an overnight scout camp. 'The kids loved it, of course. But one of the leaders was scared of moths – properly frightened. So I showed her a Buff-tip and placed it on her finger. As she cradled it, she wondered out loud why she had been scared of moths all her life.'

Attractive though we found them, the Southern Chestnuts provided no match for such nailed-on crowd-pleasers. But the story of this auburn enigma, if anything, is just as remarkable. Is it an under-the-radar native speaking of nature's secrets or a recent colonist and harbinger of a changing climate? A genuinely rare species worthy of conservation concern or a moth bucking the trend of autumn-flyers by expanding its range and numbers?

For Will, Southern Chestnut makes for a memorable final trip of the year. To celebrate, we take a final swig of Chris Fox's cider and crumple into the car.

In late November, at the cusp of autumn and winter, dusk falls early so university student Jack Morris and I must make haste to an isolated autumn-bronze woodlot in south-east Suffolk. The Suffolk Wildlife Trust cares for Groton Wood, recognising its value as remnant wildwood. Its fame derives from its stand of Small-leaved Lime, unusual in East Anglia. I know this wood from a late-summer orchid sortie, a stroll and loll in admiration of the willowy Violet Helleborine. But we owe our presence today to another tree, the Field Maple.

John Clare wrote of the 'sweet clothing of the maple tree', and I gaze upwards at its spry form, the bark scaly and ridged, the bare lower branches succeeding to bushy, golden Klimtness. The sticky floor swishes with my passage through fallen leaves. Each one is a heavily indented, sparingly veined Canada flag. Most leaves are buttercup-yellow, but others are tarnished with tin or enflamed with russet. Some are discontent to have just one colour, however impressive it may be, and exhibit the lot. Unrestrained autumn, in a leaf.

The moth we are after – Plumed Prominent – knows no life without Field Maple. It is not a moth you bump into and not one you will catch in your garden. It occurs very sparingly in southern England, in woodlands and along what's left of our hedgerows, but always with Maple and always, for reasons unknown, on chalk. Yet it is an enigma, for even where those conditions coincide, its occurrence is no given. Butterfly Conservation's Sharon Hearle has scoured suitable woodlands near her west Suffolk home without getting a sniff of Plumed Prominent. To see it, she has resorted to joining the Suffolk Moth Survey's annual evening at Groton Woods, where the moth eluded detection until the 1970s but is now well known. And she has extended the invitation to us.

As dusk thickens, Sharon is buzzing. Her autumn has been chilly and wet, furnishing scant opportunities to see moths. To stave off cabin fever, she has resorted to perusing fungi. But our lights are on and the early evening is mild, so she is happy. We are too.

'These are perfect conditions,' says Neil Sherman. Suffolk's county recorder for moths is tonight's host. 'We should catch a dozen, and they should be in quite quickly – the first about an hour after dark.'

Like Sharon, both Jack and I have endured a month of moth starvation, with no generator-toting trips into the wilds to alleviate garden fare. We are elated to see Feathered Thorns fluttering around, Chestnuts darting and male Scarce Umbers biding their time. We stroll through flurrying flakes – not snow, but male Winter Moths eddying on their hunt for love. This bodes well. We scrutinise trunks in the hope of a female Scarce Umber. Blessed with honed teenage vision, Jack swiftly spots one – the first of her gender that I have seen. She is flightless, a chunky ladybird-larva half the size of my little fingernail with stumpy excuses for wings. She crawls to a cleft in the bark at my shoulder height; I marvel at the hike that brought her here.

A few inches to the Scarce Umber's left is an even smaller flightless female moth. Moreover, this lady Winter Moth is conjoined with a winged male. I indulge in voyeurism: this is another species for which I have never encountered a female in the field, let alone witnessed copulation. She is tiny – a woodlouse would dwarf her – wrinkly and wingless. It takes a leap of faith to accept her mothness.

With these ladies of the night, the evening has already proved worthwhile, but the best is to come. An hour after we switched on the traps, Sharon spots a buzzing in the leaf mulch beside one of my traps. We have our first Plumed Prominent.

We each admire him, Sharon particularly ardently. This is what she has been missing in her nocturnal forays around Newmarket. As the moth settles, it assumes the arched-tent form of the prominents that occasionally visited our garden this summer. But there the similarities stop. Our boy – and it is a boy – has a ferocious head of hair that merges into a fur collar; his is late-autumn attire. He is gingery with swirls of beige and bronze, but this colouration is fragile. His wings are crêpe-paper thin; the slightest brush of inattentive finger would rub the scales bald.

This would be moth enough for us, but we pull up short when our eyes reach the Plumed Prominent's antennae. These mate-seeking devices are plumes indeed – without parallel among British moths. Beautifully pectinated, they resemble avocado-shaped feathers. Head-on, this is more Bat-eared Fox than insect. The Plumed Prominent is a thing of wonder.

It is also, we discover, a thing of local abundance. One hour on, and Plumed Prominents are everywhere. All are males, and they exhibit as wide a colour spectrum as Groton's leaves. I tot up thirteen sitting sedately on the outside of one trap, with others slouching on the leaf litter below and more inside. By the time we pack up, we have counted eighty-four between us. Only once in this annual survey, Neil says, have more been caught. Moreover, all are freshly emerged – probably hatched yesterday, he thinks. We've timed it perfectly.

Mission more than accomplished, we leave the Plumed Prominents be. Were we to stay later – much later – a female might come in; they are tardy types. But it's the males' antennae that I wanted to gawp at. So we dismantle our traps and climb into the car, astonished at our fortune. It is not even 8 pm. We might be home by half-nine – about the hour we were setting up traps in mid-summer. There's much to be said for autumn mothing.

It is every moth-er's dream – every naturalist's dream, indeed – to discover a species new to their country. This is unchartered territory – the entomological equivalent of the mother of all journalistic scoops, of a mountaineer scaling a hitherto inaccessible crag. Some fortunate souls I meet on my travels boast such claims to fame – Sean Clancy, Martin Cade, Nigel Jarman and David Walker among them. Almost without exception, the finder is an experienced moth-er. Recognising a species that isn't 'in the book' demands, at a minimum, knowing that book inside out and then being aware of where else to look in order to elicit an identification. Either that or experience of moths overseas. Either way, finding a new species for the UK isn't for eternal novices like me.

But it was for the man who got me into mothing. One Monday lunchtime in November 2011, James Hunter popped round the corner to see a strange moth caught by friend Norman Winterman in his urban Dartford garden. Norm tentatively called it as Red-headed Chestnut, a rare migrant. James cautiously agreed. But it didn't look right. Posting images on the internet prompted rapid-fire messages from experienced moth-ers who recognised it as Black-spotted Chestnut, previously unknown from the UK. This transpired to be on their radar as a candidate British moth because it had recently spread rapidly through northern France into the Low Countries, where it had swiftly become the commonest winter-flying moth.

Norm and James knew none of this, each having only returned to mothing the previous year after more than a decade's hiatus. What they did know, however, was that a score of moth-twitchers were chomping at the bit to see the new arrival. Excitement wasn't the word. James arranged for a viewing in his garden – less than 300 metres from Norm's – the following evening. At which point, lightning struck twice.

As one visitor, Paul Chapman, was standing in James's garden, sipping a calming cuppa after viewing the rarity, his gaze was caught by a moth buzzing around James's illuminated trap. 'Get a pot! Get a pot!' Paul screamed frantically, catching the moth in his hands. The UK's second Black-spotted Chestnut had arrived. Every jaw dropped. Two friends, two gardens, two nights, two moths. The coincidence suggested something was up. The chance of a brace of vagrants arriving nigh on simultaneously at nearly the same place, amid dense urban sprawl inland, was negligible. 'Instead,' James remembers, 'we reckoned that a female had actually arrived the previous year – and these were her offspring.' This wasn't merely a new moth for the UK but a new breeding population. Moths are like that: always changing, always surprising.

Although I wasn't into mothing in 2011, I remember keenly James's astonishment and the resultant buzz of excitement across social media. I congratulated my friend on his discovery, but it didn't even cross my mind to scoot the nine miles from south-east London to take a look myself. By late 2018, now four years into living in Norfolk and a fledgling moth-er, I was regretting this. The winter-flying Black-spotted Chestnut was firmly established in north-west Kent and now hinting at developing an outpost fifty miles north-west in Bedfordshire. James was proving a magnet for this particular moth, having caught forty himself – perhaps two-fifths of all British records. Black-spotted Chestnut even followed James to his new garden after he moved house. But there was no suggestion of 'Hunter's moth' mooching towards Norfolk any time soon. I wouldn't be catching it in my garden.

And yet I really, really wanted to see it. Whether out of homage to James or due to inexplicable obsession, this was one moth I simply *had* to see during my year-long quest. Even over Kentish Glory, Barberry Carpet or Clifden Nonpareil. This unremarkable pebble-grey moth may have been the winged equivalent of an actuary, but missing it was not an option.

I flunk my first chance. In January, right at the start of my quest, James alerts me that he has caught one in his garden. But family commitments preclude driving to see it. He catches another in early November and again phones. I am conflicted. I really want to see Black-spotted Chestnut but, after a year of catching my own moths, have become less enamoured by 'pot-twitching'. A guaranteed tick, for sure, but devoid of the elation of seeing one resting innocuously in one's own trap. Teeth gritted, I elect to bide my time until a suitable opportunity arises to yank the cord on my generator for what will surely be the final time this year. I know Sean Clancy and Nigel Jarman want to catch one too. So I wait.

I receive a text on the final Saturday of November, just as I am taking my daughter swimming. 'We're on for tomorrow night,' Sean writes. 'We're trying a new woodland – never been surveyed for moths before – south-east of Dartford.' He has no idea whether Black-spotted Chestnut would occur there, but it seems plausible. Sean's a man to push the boundaries of knowledge, to try new places, and he's a man whose instincts I trust. If he reckons there's a sniff, I need to be there.

The next evening I pull up in an ominously secluded car park where Sean is waiting with fellow Kent mothers Nigel Jarman and Julian Russell. It is mild enough for

Sean to be wearing cut-off shorts. While we're setting up, a giant of a man – all shaved head, don't-mess-with-me beard and muscular heft – emerges from a van to unleash possibly the scariest-looking mutt I have ever seen, which promptly pisses on one of Nigel's traps. Goliath looks towards Sean's nearest trap, which sits atop a white sheet in the middle of a grassy area by the car park.

'What the hell are you guys doing?' he asks, voice thick with both Polish accent and confusion. Our explanation merely deepens Goliath's bewilderment. 'Moths? What are *moths*?' It is unclear whether the English word is unfamiliar or the animals themselves. Sean covers both bases by showing the man a copy of the identification guide he wrote to UK moths, Sean's self-penned Bible. Goliath shows little recognition, but a smidgeon more curiosity.

'Have it,' Sean says generously. 'Take the book.' Now completely puzzled, Goliath does as he is told. And a smile spreads across his massive face. These unfathomable strangers have brought gifts. Sean's work here is done, and Goliath leaves us clear to wander into the woods.

Birch trees salute the claggy paths, their bushy blow of golden leaves brightening the dusk like the final flourish of glittering fireworks. Deeper into the wood grow Field Maples – but there will be no Plumed Prominent here, not on this clay – and a hint of oak. An aeroplane roars low overhead, descending towards nearby London City Airport. In the wilds, we are not.

James Hunter arrives. How could I not extend Sean's invite to the man who knows Black-spotted Chestnut better than any other Brit? Phil Saunders also strolls up, back from Dorset for a weekend in his native Dartford.

We scatter nine traps in clearings spiked with wooden exercise equipment. Sean's local woodland is no ancient-forest reserve but a multi-use site encouraging urbanites to temporarily relinquish the materialist mayhem of the nearby massive out-of-town shopping centre for a semblance of fresh air. No wonder nobody has bothered surveying moths here before. I am not hopeful.

Yet my grizzling is misplaced. Within a couple of hours, we have totted up twenty-one species – a highly respectable total anywhere in Britain for so late in the year. The roll call speaks with simplicity of seasonal shift. Among the November Moths are Autumnal Moths; December Moths are everywhere. There are Winter Moths and Northern Winter Moths. We get the picture: the year's end is fast approaching. But there is no Black-spotted Chestnut. The last trip of the year is doomed to failure. After repeated highs, I will end on a low.

After the 6 pm trap check, Phil leaves; the road to Dorset is long. James, our lucky mascot, also departs, pleading family chores. He promises to phone should his garden trap miraculously contain a Black-spotted Chestnut: we call the option 'Plan H'. But James is not optimistic. He catches 'between zero and five in November and December, and I've already caught one'. I do the maths. James traps every night. That suggests, at best, that Plan H provides a one-in-twelve chance of success tonight. My mood blackens like the markings on our target's wings.

Just after 8 pm, our remaining gang of four chivvy ourselves into another trap round. We agree to pack up at nine; these next twenty minutes may be our last roll of the dice.

Before we've stumbled ten paces into the pitch, my phone rings. It's James Hunter. My heart pulses. His Dartford twang is music to my ears.'You're not going to believe this, mate, but I've got one. I've caught you a Black-spotted Chestnut.'

Punching the air, I giggle hysterically and blurt out the news to the others:'James has got one! He's got one! I don't frigging believe it.' Our collective mood jolts into incandescence. High-fives and tension-defusing laughter all round. Plan H has come up trumps. We agree to do a final count of moths, pack up our traps and make haste to James's garden. It may be a pot twitch, but it's on the way home. Furthermore, we've tried to catch our own, so this reward is earned.

Just over an hour later we trip through the dark along a potholed track to James's house. There are no streetlights here – a curse for those who see them as providing security, but a blessing for moth-ers who need dark skies. Slipping down the side of the building, we edge carefully round the actinic light that occupies the narrow gap between house wall and garden fence. We rejoin James, cramming into his kitchen. Teenagers lurk in the next room, watching television, ignoring their dad and his late-evening visitors.

'You don't know how lucky you are,' James says. 'It wasn't on the egg-trays or the wall where they usually sit. It was hiding under a wooden strut behind the trap. That's snatching victory from the jaws of defeat, that is.'

He reaches into the fridge, extracts a pot and passes it to me. My hands are shaking as I tap the moth onto a neatly cut leaf.

In all honesty, this moth is not a looker. It is dauby clay-brown, suffused with grey. Prominent veins traverse the black pawprints that informed its English name. Nor, eight years after its discovery in the UK, is the Black-spotted Chestnut particularly rare. Anyone who has wanted to see one should have done so by now. And yet it is perhaps the most personally resonant moth of the year.

My moth journey started in July 2012 with a female Poplar Hawk-moth that James Hunter extracted from a scruffy rucksack and thrust upon me. Without Hunter's hawk-moth – without James's persistent attempts to convert me – the moth odyssey that dominated this year would never have happened. That the year would end with Hunter's chestnut must have been written in the stars. I shake James's hand. Inside, I am welling up.

I bid everyone farewell, and drive home through the night for the final time this year. Norwich has never seemed so far away. For what feels like the hundredth such occasion in twelve months, I finally pull up in our driveway during the witching hours.

This year, I have travelled 14,000 miles in pursuit of Britain's rare and remarkable moths. I have hiked up mountains, waded through mires and explored forests. I have enjoyed moths at 139 locations across twenty-seven counties or equivalent. My notebooks contain 396 entries over 258 days. I have seen 1,135 different types of an insect that I once deemed too boring to bother with – and loved every single one of them. I have been bewitched by creatures I never knew existed and learned things about them that I would never have believed possible. I have spent time with scores of people dedicated to studying or saving wildlife, often in the most trying of circumstances. I have slept rough a dozen times and missed sleep entirely on many more nights. I have traumatised my body clock, wrecked my car and sometimes disappointed as both husband and father. But I have loved every minute of it. There is, indeed, much ado about mothing.

Epilogue

The day after the year's closing moth trip, the charity Butterfly Conservation publishes a remarkable book. Collating 25.6 million records of larger moths in Britain and Ireland that span the 275 years up to 2016 is a mighty undertaking. Even more so the subsequent analysis of that data to map the distribution of nearly 900 species and to determine population trends spanning nearly half a century. But the *Atlas of Britain and Ireland's Larger Moths* goes further still, interrogating the mountain range of information to assess the overall status of moths in the two countries. Their conclusions are unsettling.

The status of roughly one in eight resident macromoths – more than a hundred species – has become so parlous that they are categorised as nationally threatened or nearly so. For species where long-term abundance trends in Britain could be determined, nearly two-thirds have declined – with one-third exhibiting statistically significant decreases, many dropping by more than half. At first blush, this suggests Britain has not been spared the so-called global insect apocalypse that has spawned newspaper headlines in recent years. However, the *Atlas* reveals a more nuanced picture. Some moths are actually dramatically more common: by 2016, Buff Footman – a routine fixture in our summer garden – has become 845 times more abundant than in 1970. In terms of distribution over the same period, slightly more moths have seen a statistically significant increase in distribution than those whose range is shrinking or showing no significant shift either way.

Britain's moths are in continual flux – even at the binary level of presence or absence. At the end of each of the past three decades, one of the *Atlas* authors, Mark Parsons, has assessed moth comings and goings. His most recent instalment was published a few months after the *Atlas*. Since 1900, Mark calculates that the UK has gained nearly 140 colonists – most since 1990, their arrival expedited by factors such as the horticultural trade and climate change. By and large, their appearance is welcome or at least anodyne.

But there's an unsavoury side to the UK's moth balance sheet. Mark reckons that fifty-two species have probably gone extinct here. Their ilk inspired Sir Harrison Birtwhistle to compose *Moth Requiem*, a 2012 choral work starring twelve female vocalists who incant Latin names of the extinct. Mark's latest tentative propositions for the litany of gone moths include two that I sought during my year-long quest: the heathland-dwelling Speckled Footman and littoral *Scythris siccella*. On the plus side, perhaps fifteen species previously judged extinct have recolonised or been rediscovered – including Kent's scintillating *Hypercallia citrinalis* in summer 2019 – and Speckled Footman itself was subsequently refound in 2020.

Change, then, is a constant in British mothery. The *Atlas* summarises the drivers of this incessant reshaping. Intensive approaches to land management, particularly on farmland, harm moths. So too large-scale replacement of wildlife-rich landscapes such as ancient woodland or chalky grassland with concrete, conifers or carrots.

Climate change, however, plays both ways. Most species seem likely to benefit – as one might intuit for generally warmth-loving animals. But species that need colder winters – Mottled Umber among my garden denizens, for example – are suffering dramatically. Others may be retreating uphill or being pushed northwards. Unless we arrest the climate crisis, montane species and Scottish exclusives will eventually have no place left to go. Their brink beckons.

Alongside such inadvertent, largely negative transformation, I have been fortunate to witness much directive, positive change throughout my year of travels. I have observed many individuals refusing to be beaten by the prospect of lepidopteran extirpation, instead striving to improve the lot of moths. Mark Parsons flipping Dorset scree to count the grains of rice known as *Eudarcia richardsoni*. Gaby Flinn leading Cairngorm volunteers to Kentish Glory. Fiona Haynes inspiring volunteers to plant Barberry hedgerows. The late Douglas Boyes devoting nights to Wytham Woods and light-struck caterpillars. Jen Nightingale ascending scary slopes to survey Silky Wave. Dave Grundy, mothing far from home for 200 nights a year, irrepressibly pushing the boundaries of knowledge. Sharon Hearle persuading highways authorities to rotavate Breckland road verges in the name of Grey Carpet. Steve Lane enthused to pursue a brighter future for Norfolk's Marsh Carpet. Neil Ravenscroft returning to the cause of New Forest Burnet after two decades away. Brian Hancock beavering away for Lancashire's remaining Netted Carpets. Sean Clancy battling for Sussex Emerald and Marsh Mallow Moth, headliners among Kent's Magnificent Moths.

Abby Gibbs and Chris Thain devoting a third of their nights to understanding moths around Poole Harbour. Zoë Ringwood making Essex a safer place for Fisher's Estuarine Moth. All doing what it takes to make their bit of the world a better place for nature.

Change is manifest in me too. Although not quite the moth's revelatory transformation from egg to winged imago via caterpillar and pupa, mine has been a year like no other. I have been awed by the ways in which moths matter, from their service as pollinators to their sufferance as prey – and perhaps even to their resolution of plastics. I have been gobsmacked to encounter moths that produce sounds, jam bats' radars, imitate twigs or become one with lichen. While the rapture of intoxication started at home, the voyages of discovery offered by moths opened my eyes to wondrous places of which I was blithely unaware. Through such varied experiences, I have learned to furnish my own explanatory narratives to successive real-life natural-history unit productions. A kaleidoscope of moths has coaxed me from my comfort zone and, in times of mental or emotional hurt, become my balm. Through them, I have grasped how to become – and stay – present in nature, now fixing my attention on the current and palpable without succumbing to digression or distraction. A completer-finisher by habit, I have nevertheless come to embrace the rich infiniteness of moths. But, equally, never have I felt so compelled to thrust onwards, to seek afresh, to push the boundaries of knowledge, to contribute.

I should not be surprised that moths have changed me. Nobody is born a moth-er, but anybody can easily become one. Throughout Britain, I see how moths have enhanced people's lives – folk who have moved along the curve from hatred of pests or fear of unpredictable flappers to positivity, wonder, even love. My former neighbour Estelle Alwan and family professing to be 'converted' by the loveliness of moths. Mark Youles pursuing 'Deathies' around his kitchen. Mary Laing swopping musical scores for moth traps upon retirement. Another retiree, Dorothy Beck, granting her Kent garden wildness to help Sussex Emerald. James Hunter, resolutely trapping repeatedly alone, yet actually yearning for company. And Maya, with the Poplar Hawk-moth – the moth that changed me – pooing on her face during her birthday party. The moth that changed my life has now also left an indelible impression on my daughter.

Further reading

I consulted hundreds of books, scientific papers, magazine articles, newspaper reports and websites during research for this book, but here are some suggestions for further reading that may deepen your appreciation of moths and their environments.

Allan, P.B.M. (1937) *A Moth-Hunter's Gossip*. Philip Allan & Co. Ltd.

Clancy, S., Top-Jensen, M. & Fibiger, M. (2012) *Moths of Great Britain and Ireland*. Bugbook Publishing.

Ford, E.B. (1967) *Moths*, 2nd edn. Collins.

Gandy, M. (2016) *Moth*. Reaktion Books.

Lees, D.C. & Zilli, A. (2019) *Moths: Their Biology, Diversity and Evolution*. Natural History Museum.

Leverton, R. (2001) *Enjoying Moths.* T. & A.D. Poyser.

Lowen, J. (2016) *A Summer of British Wildlife.* Bradt Travel Guides.

— (2021) *British Moths: A Gateway Guide*. Bloomsbury Wildlife.

Majerus, M. (2002) *Moths*. Harper Collins.

Manley, C. (2015) *British Moths: A Photographic Guide to the Moths of Britain and Ireland.* 2nd edn. Bloomsbury Wildlife.

Marren, P. (2019) *Emperors, Admirals & Chimney Sweepers*. Little Toller.

McCarthy, M. (2015) *The Moth Snowstorm*. John Murray.

Newland, D., Still, R. & Swash, A. (2013) *Britain's Day-flying Moths*. WILDGuides.

Randle, Z., Evans-Hill, L.J., Parsons, M.S., Tyner, A., Bourn, N.A.D., Davis, T., Dennis, E.B., O'Donnell, M., Prescott, T., Tsordoff, G.M. & Fox, R. (2019) *Atlas of Britain and Ireland's Larger Moths.* Pisces Publications.

Salmon, M.A. (2000) *The Aurelian Legacy.* Harley Books.

Salmon, M. & Edwards, P.J. (2005) *The Aurelian's Fireside Companion*. Paphia Publishing.

Skinner, B. (1998) *Moths of the British Isles*, 2nd edn. Viking Books.

South, R. (1961) *The Moths of the British Isles,* series I and II. Frederick Warne & Co.

Sterling, P. & Parsons, M. (2018) *Field Guide to the Micromoths of Great Britain and Ireland*. Bloomsbury Wildlife.

Waring, P. & Townsend, M. (2017) *Field Guide to the Moths of Great Britain and Ireland.* 3rd edn. Bloomsbury Wildlife.
Young, M. (1997) *The Natural History of Moths.* T. & A.D. Poyser.

Here are the other books from which I have drawn:

Barkham, P. (2010) *The Butterfly Isles*. Granta.
Bersweden, L. (2018) *The Orchid Hunter*. Short Books.
Cocker, M. (2018) *Our Place*. Jonathan Cape.
Donne, J. (1633) 'The Sun Rising'. In: *Collected Poetry*. Penguin.
Dunn, J. (2018) *Orchid Summer*. Bloomsbury.
Elder, C. (2015) *Few and Far Between*. Bloomsbury.
Fowles, J. (1987) *The Blinded Eye*. In: Mabey, R., ed. *Second Nature*. Jonathan Cape.
Heine, H. (1851) *Waldeinsamkeit*. In: *Gedichte*. Diogenes Verlag AG.
Hudson, W. H. (1919) *The Book of a Naturalist*. J.M. Dent & Sons.
Keats, J. (1819) 'Ode on Melancholy'. In: *Selected Poems*. Penguin.
Kerridge, R. (2014) *Cold Blood*. Vintage.
Marren, P. (2018) *Chasing the Ghost*. Square Peg.
Muensterberger, W. (2014) *Collecting: An Unruly Passion*. Princeton University Press.
Nabokov, V. (1951) *Speak, Pemory*. Penguin.
Savill, P., Perrins, C., Kirby, K. & Fisher, N. (2010) *Wytham Woods: Oxford's Ecological Laboratory.* Oxford University Press.
Sjöberg, F. (2015) *The Fly Trap*. Penguin.
Taylor, M. (2013) *Dragonflight*. Bloomsbury.
Wheeler, J. (2017) *Micro Moth Vernacular Names*: Clifton and Wheeler.
Woolf, V. (1932) 'Reading'. In: *The Common Reader*. Hogarth Press.
— (1932) 'Four Figures', op. cit.
— (1932) 'The Death of the Moth', op. cit.

Finally, there are some excellent internet resources, organisations and regular publications to guide moth-ers of all abilities:

Atropos magazine (www.atropos.info)
The Entomologist's Record and Journal of Variation (www.entrecord.com)
Butterfly Conservation (https://butterfly-conservation.org/moths)
Moths Ireland (www.mothsireland.org)
National Moth Night (www.mothnight.info)
National Moth Recording Scheme (www.mothscount.org)
UK moths website (www.ukmoths.org.uk)

Acknowledgements

Research for this book was conducted with the help of a grant from the Author's Foundation of the Society of Authors, for which I am most grateful. At Bloomsbury, I thank Jim Martin and Alice Ward for making the book what it is. Thanks also to Charlotte Atyeo and David Hawkins for their input on the text. Jasmine Parker designed the lovely cover. The book would not have been possible at all without the support and love of my she-moth and caterpillar – Sharon and Maya Lowen, respectively. I have never been so negligent of what actually matters in life. That I have been neither divorced nor disowned is testament as much to my girls' tolerant understanding as to our emotionally robust tethering. But I learned to play my part too, realising the importance of interspersing my explorations aboard *Le Bateau ivre* with regular anchorage in the safe port of family. Nor would this book have happened without the companionship and moth nous of Will 'Wingman' Soar or the initial introduction to moths (and ongoing companionship and advice) from James Hunter. Sharon Lowen, Sarah Morrison and Will Soar kindly read the draft text in its entirety, providing many useful comments.

A huge number of individuals and institutions provided advice, shared tales, authorised mothing trips, participated in adventures and/or commented on elements of the text. With no apologies for the length of this list, I am grateful to: @BritishMoths (on Twitter), Nick Acheson, Mick A'Court, Ian Alexander, Alison and

Chris Allen, the Alwan family (Ali, Estelle, Matthew and Sienna), Dave Andrews, Mark Andrews, Dave Appleton, *Atropos* magazine, Back from the Brink, Chris Balchin, Barry Banson, Andy Banthorpe, the late David Barbour, Patrick and John Barkham, Iain Barr, Richard Bateman, Bawdsey Hall, Dorothy Beck, Matt Blissett, Bernard Boothroyd, Evan Bowen-Jones, Neil Bowman, Claire Boyes, the late Douglas Boyes, Bristol Zoological Society, British Entomological and Natural History Society, Dave Broadfoot, Gareth Brookfield, David Brown, Joe Burman, Burnet Study Group, Stuart Butchart, Butterfly Conservation, Zoe Caals, Martin Cade, Lucy Carden, Bex Cartwright, Martin Casemore, Matt Casey, Sarah Cassidy, Paul Chapman, Lucia Chmurova, Sean Clancy, Jon Clifton (Anglian Lepidopterist Supplies), Mark Cocker, Lisa Cox, Terry Crawford, Mischa Cross, Jon Curson, Simon Curson, Mark Davison, Tony Davison, Matthew Deans, Dorset Wildlife Trust, Ann and Andrew Duff, Dungeness Bird Observatory, Jon Dunn, Guy Dutson, Craig Edwards, Terry Elborn, Gary Elton, Environment Agency, Annie and Justin Farthing, Sue Fellowes, Gabrielle Flinn, Sean Foote, Forestry England, Martin Fowlie, Chris Fox, Steve French, Steve Gale, Adrian Gardiner, Abby Gibbs, Mark Grantham, David and Penny Green, Nick and Sian Green, Brian Hancock, James Harding-Morris, Kate Harris, Robin Harvey, Tom Hayek, Fiona Haynes, Sharon Hearle, Max Hellicar, Alexander Henderson, David Hermon, Josie Hewitt, Gary Hibberd, Martin Hallam, Malcolm Hillier, David Hipperson, Ben Hoare, Pauline Hogg, Gill Hollamby, Dave Holman, John Hooson, Ian Hunter, James Hunter, Nigel Jarman, Clive Jones, Graham Jones, Josh Jones, Ade

Jupp, Kent Wildlife Trust, David Knight, Mary Laing, Sophie Lake, Rupert Lancaster, Steve Lane, Andy Lawson, Marcus Lawson, Richard Lewington, Ben Lewis, Owen Leyshon, Durwyn Liley, Pili López, Alison Lowe, Tom Lowe, Jenny and David Lowen, Amy Lowen, Chris Manley, Pete Marsh, Richard Mason, Steve Masters, Dougal McNeill, Peter Moore, the Morris family (Jack, Josh, Nina and Pete), Sarah Morrison, Moths of Poole Harbour, Andy Musgrove, Joe Myers, Steve Nash, Natural England, National Trust, Ross Newham, Lynnette Nicholson, Jen Nightingale, David Norgate, members of the Norwich and Norfolk moths WhatsApp group, Ashen Oleander, Chris Panter, Henry Page, Steve Palmer, Ed Parnell, Mark Parsons, Pensychnant Conservation Centre, Steve Peters, Luke Phillips, the Phillips family (Adrian, Monika, Kitty and Matthew), Stuart Piner, Portland Bird Observatory, Tom Prescott, Rare Invertebrates in Cairngorms project (a partnership between Buglife, Butterfly Conservation, Cairngorms National Park, RSPB and Scottish Natural Heritage), Zoë Randle, Neil Ravenscroft, Catriona Reid, Ian Rickards, Zoë Ringwood, Ian Roberts, Pete Robertson, Ian Robinson, Helen Rowe, RSPB, Julian Russell, Sandwich Bay Bird Observatory, Ken Saul and members of the Norfolk Moth Survey, Debby Saunders, Phil Saunders, Scottish Natural Heritage, Harry Scott, Matt Shardlow, Dave Shenton, Neil Sherman, Tim Sievers, Alick Simmons, Alan Skeates, Murray Smith, Will Soar, Adrian Spalding, Christine Steen, Phil Sterling, Peter Stronach, Suffolk Moth Group, James Symonds, Andy Taylor, Darren Taylor, Stewart Taylor, Mark Telfer, Chris Thain, Kelly Thomas, Julian Thompson, George Tordoff,

Mark Tunmore, Jacques Turner-Moss, University of East Anglia, University of Oxford, Nigel Voaden, Martin Wain, Dave Wainwright, David Walker, Mike Wall, Steffan Walton, Paul Waring, Jill Warwick, Nick Watmough, Mike Watson, Steve Wheatley, Jim Wheeler (Norfolk Moths), Paul Wheeler, Graham White, Steve Whitehouse, Chris Williams, Karen Woolley, Leon Woodrow, Stewart Wright, Mark Young and Mark Youles. My apologies to anyone inadvertently omitted.

Index